EPANET and Development

How to calculate water networks by computer

Second Edition, Revised and Expanded
April 2021

Santiago Arnalich

EPANET and Development

How to calculate water networks by computer

Second Edition, Revised and Expanded
April 2021

ISBN: 978-84-613-1477-5

Cover photo: Damage from the 2004 Tsunami in Meulaboh, Indonesia
Errata at: <u>www.arnalich.com/dwnl/xlipacoen.doc</u>

Translation and proofreading: Amelia Jimenez, Peter Cruddas, Mary Brown, Maxim Fortin, John Butterworth, Binyam Bedaso, Cornelius Mpesi and Tinotenda Muvuti.

FOREWORD

For those of us accustomed to open a tap and see the water flowing, it may seem strange that we still need to publish, review, and translate books about water distribution networks, because it seems so easy to have tap-water. In my first mission as a WASH engineer, I organized an excursion on a sunny Sunday to explain to my colleagues from the International Committee of the Red Cross how the water supply they were using on a daily basis is actually functioning: we visited a catchment, followed the pipeline, a pumping station, visited a water treatment plant and a large reservoir situated on a hill above the city. They all told me that they did not know how water was flowing from the tap until then, but that they would remember how it works.

Some decades later, the water supply situation has improved in many places, and humanitarian actors increasingly dare to repair existing water infrastructures damaged by conflicts or natural disasters instead of wasting energy and money in building expensive temporary water supply systems such as water trucking and other typical humanitarian Do-It-Yourself solutions. But to build, use, repair or improve an existing water supply system, it's crucial to understand its behavior, and hydraulic calculations are a must to avoid spending time and money building a system in which the water is unable to compete with gravity or friction forces! Otherwise, the miracle of water flowing from the tap would just not happen… Santiago Arnalich has decided that EPANET, a freeware developed by the US Environmental Protection Agency, potentially makes hydraulic calculations accessible to every engineer if they benefit from a good introduction in the subject. This is just one book of his series of technical manuals which allows WASH engineers to map the site in which they intend to intervene, design and calculate gravity-fed systems, equip a borehole, select the right type and then maintain generators or design a water distribution system from where water will flow.

With climate change increasingly affecting the water cycle, investments in good infrastructure such as centralized water distribution remains a very efficient way of delivering the first food of humanity, which is … water. But there is also a need to manage water resources more carefully to avoid wastage of water or energy. This can only be done by complementing a water distribution system with an adequate water resource monitoring system. Long-term measurements of yields of springs or of static groundwater levels are needed to assess the evolution trends of water resources. On the other hand, measurements of water quantities injected in water system and precise evaluation of water losses allow for fast repairs as soon as critical levels of water losses are reached. Sensors and water meters installed at critical points in a water supply system allow today a constant monitoring of its behavior and early interventions either at the level of the water resources (e.g. by reducing the pumping hours if the water table is dropping) or in the management of the distribution network (e.g. by distributing water only some hours per day or by reducing the overall pressure until repairs to reduce water losses are finalized).

If we want the miracle of water flowing through the tap to last for many more years, we'll have to integrate the management of water distribution system in a larger perspective encompassing integrated water resource management from the source to the end user. Once the water is used (which often happens a couple of decimeters away from the tap where it flowed out clean and safe), wastewater needs then to be collected and treated so that it can be recycled in the global water cycle.

With his book, Santiago Arnalich paves the way to better design of water supply networks, which will save money, time and energy. We are proud to be associated in such a noble endeavor and hope that this book will help keep the water flowing.

Marc-André Bünzli
Humanitarian WASH thematic advisor
Swiss Agency for development and Cooperation

Contents

CHAPTER 1

Introduction

Some important points before you get started

• This manual is based on a personal view. Things are not absolute. Many decisions during the design process are debatable. Please, keep a critical mind and remember that what I present here is not "the way to design a network", but just one of them.

• Practice is important therefore we recommend the exercise manual *EPANET and Development: A progressive 44 exercise workbook* available at:
www.arnalich.com/en/books.html.

• This manual has been created for use in development projects and therefore some recommendations may not be suitable for use in developed countries.

• This manual is limited to the design of a network. Analysis of existing networks needs a more complicated approach and more elaborate techniques.

• There are not many people who work in development full-time for a long period, and as such this is not intended to be a cutting-edge manual, but one that you can return to when you need it.

• This book is aimed at networks with a minimum size. The peak coefficient approach used is not suitable for very small networks or indoor installations. To avoid problems, and for other equally important reasons (extensions, fires, etc.), try not to install pipes smaller than 63-75 mm with the only exception of house service connections. EPANET cannot work directly using simultaneity coefficients because the conservation of mass laws would be violated.

▶ https://youtu.be/Jt6nGTZ5CgE   (Choosing a design Flow method)

• If you find any errors, please let us know at: publicaciones@arnalich.com

Once you have read this manual and with a little practice you will:

1. Be familiar with basic aspects of EPANET.
2. Be able to compile the information you need to get started.
3. Be able to design new networks.
4. Be introduced to techniques of repair, extension and optimization of existing networks.

New design vs. existing networks

This manual covers the design of new, isolated networks and is the first step to the analysis of existing networks which is more complex. Design may seem more intimidating at first but networks that are already built add the following difficulties:

1. Problems with **data reliability** in development work are very common. Plans, calculations and up to date information are often missing. In new networks, the data is 100% reliable.

2. The **location and size of each one of the leaks** are unknown. New networks are tested under pressure when they are half buried so leaks can be detected

3. The **diameter and roughness of the pipes have been modified** due to scaling and oxide deposits, increasing the flow resistance greatly.

4. Normally **people do not know the condition of the network**, which valves are not working or which ones are blocked in intermediate positions.

The process of upgrading an existing network using EPANET is complicated and expensive. It requires field measurements, and frequently, expensive and specialized equipment; due to lack of resources, the calibration is rarely done. Maybe the practical approach is to locate and uncover the most common problems, such as blocked valves and disconnected pipes.

When should I build a network?

Networks are expensive and need considerable organizational capacity. The basic infrastructure needs very expensive components and once they are neglected it is very difficult to repair them back into shape. However, under certain common conditions, there are no alternatives that will supply such a great amount of water at such a low cost. The network is then the best option. These conditions are the following:

1. **The population is concentrated** or has the potential to increase in existing occupied areas. Less piping is needed to cover more people, making it more cost-effective.

2. **The population has social cohesion** with institutions capable of dealing with management of the network and proper structure which guarantees responsibility will be taken. Nomadic populations are not good candidates as frequently, there is no ownership of the systems which ends up being a "source of income" to be exploited. On the other hand, refugee camps, which have strong international support and designated people responsible for the water systems are, generally, ideal candidates for installing a network.

3. **The water source can be exploited in a sustainable way.** The creation of a network encourages water consumption because it reduces one of the limiting factors for consumption which is water transport. The water source should be able to handle this increase in demand. An area served by a spring or an aquifer which shows clear signs of overexploitation is a bad candidate for a network.

4. **It should not create environmental problems,** especially stagnant water. The water that is obtained and transported by a network has to go somewhere, therefore requires drainage. If the land is very flat, soil impermeable or there is no sewage system, the network will create more problems than solutions.

5. **Topography**. Without excluding them as an option, networks on flat very flat and horizontal surfaces are more complicated and costly, more prone to stagnant water problems, and continual pumping through lack of adequately sized elevated tanks. There are also problems with pipes which suffer from water hammer due to air that cannot escape from high points. A network should be considered as the first option in areas where gravity systems can be installed over relatively short distances, where there is a high water source or there is a good site to build a tank.

Why should we calculate networks?

Although humanity had to wait until the year 1936 for a mathematical tool that could calculate grid networks[1] to be developed, and that my computer needs a few seconds to calculate a simplified network of Taetan (40.000 people), it is widely believed that networks will work almost by themselves or that they can be eyeballed with a couple of obsolete guidelines.

---

[1] *Cross, Hardy. Analysis of flow in networks of conduits or conductors. University of Illinois Bulletin No. 286. November 1936.*

The proliferation of people who assure you they can calculate networks in their heads is really exasperating. It is not surprising that many of those networks never work after endless rotations of "experts" and years of active involvement from donors.

Eyeballing a network is leaving a community's health, well-being and economic development to chance. Let me insist here: if you guess a network, it will be chance that determines the health, well-being and economic development of the community despite the efforts of everybody involved. Here you have more reasons to calculate networks:

1. Networks that aren't designed properly **disregard the work and effort of communities** that are asked to collaborate.

2. **They are dangerous**. The emptying and filling of pipes due to the lack of pressure sucks pathogens into the interior of the pipes, facilitating disease proliferation.

3. **They are expensive to maintain**. Depressurized networks get full of air. When they are filled by water, the air needs to be evacuated. This bleeding has to be extremely careful in order to avoid water hammer which destroys pipes and creates leaks. Some studies in developing countries where there are frequent water cuts show that pipes that are constantly filled and emptied get broken 10 times more than should be expected[2].

4. **They are rarely extendable**. Due to lack of clear objectives in the design and improvisations it is difficult to expand networks as they are built with no criteria.

5. **They are uneconomic**. They do not use the available resources rationally, either because they are oversized or expensive to operate. For example, a common measure to "solve" networks which have points lacking in pressure is to further elevate the tank from which they are fed. This results in enormously increased energy consumption due to thousands of tons of water being elevated for the sake of a few points. Worse still, they waste the time of the people who use them, squandering their productivity with unnecessary queuing and waits.

6. **They put Human Rights at risk**. They force users to adapt to the network, rather than the other way around. Thus, a large part of the social benefits is lost even if they do provide *some* water. For example, children end up missing school to collect water early in the morning, and since girls are often chosen for this task in their family, school, or community, the gender gap is widened.

---

[2] *Lambert, A., Myers, S. and Trow, S. (1998) Managing Water Leakage: Economic and technical issues. Financial Times Energy.*

In humanitarian water projects, with professionals who have to cover very broad fields and wear different hats (engineer, anthropologist, hydrogeologist, sociologist, evaluator...), it is inevitable that some of us know very little about a particular field. **Make sure that you are not caught in the grip of the Dunning-Kruger effect**, a cognitive bias in which people with little capacity for a task overestimate their capacity ostensibly. **If you can't do the design, outsource it!** But don't dismiss the need with three random ideas. Networks must be properly calculated. **If you supervise other people's projects, demand the calculations!** A lot of money would have been saved on failed interventions if donors had incorporated that into their regulations.

On the other hand, with guidance, and some patience, it's not a particularly difficult task; so, don't panic either.

Adapting to the context

There are many manuals on hydraulics, and there is no need to make any more. So then, why another manual? **Because international aid projects are not like projects in developed countries.**

Let's look at two cases:

Networks in developed countries are designed with three objectives in mind. The first objective is that the hydraulics work meaning that they are capable of meeting the water demand. Secondly, that the water delivered to the user is of a very high quality. And thirdly, that it is resilient to breakdowns, electrical failure, etc. This last objective requires considerable investment including spare equipment and lots of additional pipes, among other expenses. When other needs such as vaccination, education or lack of work are threatening a population, does it really make sense to build elaborate electrical failure-proof networks? If you are still not sure, consider the consequences of an interruption in water supply during a power cut and compare that with the consequences of not having vaccinated the children against polio.

Another interesting issue is the demand for fire protection meaning the amount of discharge and storage required to fight a fire. In developed countries, the network allows a discharge much larger than citizen's consumption and therefore the pipes and tanks that are installed are much bigger than the ones required to cover normal demand. Many poor countries have adopted Western codes and regulations but, does it make sense to over-engineer a network to provide a discharge of 32 l/s at a certain point and have tanks of 230 m^3 when the resources for fire protection are limited to the buckets that the villagers have? Would it not be more sensible to use that investment, let's say 100,000 extra USD, to aid the creation of local business through microcredits?

All these circumstances make decisions very difficult and delicate. For these decisions, the rigor of Western manuals does not help.

Introduction to network design

A network is designed by making calculations that guarantees it is going to work correctly. There are four main functioning parameters for development projects:

- **Pressure**, which will ensure that the users will receive water at every point considered.

- **Velocity** in the pipe, which determines if the proposed network is too big and expensive to build (low velocities) or too small and too expensive to operate (high velocities).

- **Age,** which is the time water spends in the network meaning the quality deteriorates as the residence time increases.

- **Chlorine concentration,** which will ensure that the water is drinkable, and will determine whether taste causes water users to reject it or not.

There are infinite solutions to supply water under different given circumstances, and many of them are viable and reasonable according to the criteria we have just mentioned. Consider, for example, the two ways of connecting water points, to the tank:

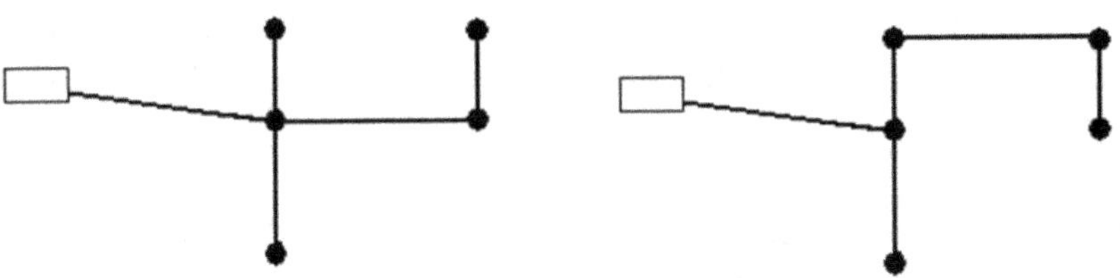

In order to choose one of them, we need to find another parameter, usually economic. Basically, it is a matter of finding the cheapest way not only for the construction of the network but also its operation costs. The method of comparing cost recovery of the initial investment against the operation of the network will be covered in chapter 8.

However, the economic criteria are not the only ones. There are other concerns that result from the overall shape of the network, one of the most important is the branched vs. grid networks.

In a **branched** design, pipes get branched off in a similar way to tree branches. Water can only get to each point through a certain route therefore there are two drawbacks, lack of reliability and quality problems due to the stagnation of water within the network. In order to solve them, networks are built with a grid layout (or a honeycomb) which although are more reliable and hygienic are also more expensive. In the image below, you can see the transformation of a branched network into a grid through the installation of pipe *t* to close the loop.

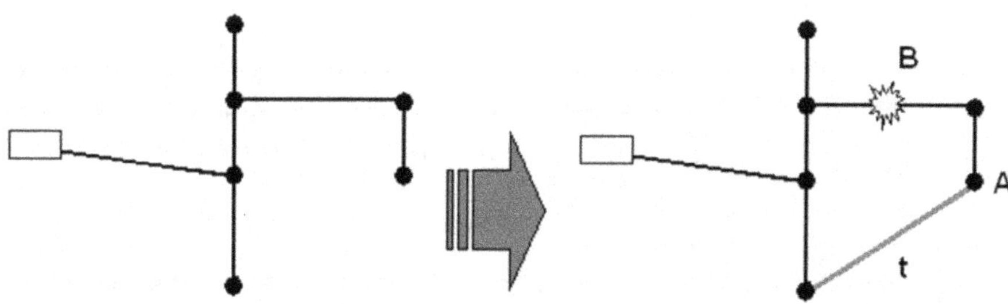

By adding pipe *t*, the water point A can receive water from both the north and from the south, so that if B breaks, the water supply in A will not be interrupted. Furthermore, water is not blocked anywhere. If not consumed in A, it will circulate to somewhere else.
Returning to the issue of trying to find the cheapest network, it is clear that we always would end up building the branched design however, the tendency is to build grids!
In a development context, there are some other reasons for grids:

- **Robustness**, especially when the material is expensive, not easily available, and the inexperienced users are dealing for the first time with a new water network and new staff.

- **Adaptability**, branched networks do not allow big modifications or extensions; it is more difficult to adapt them to keep up with the increase of population.

- **Scarcity and reliability of data** recommends grids because the branched designs need more precise knowledge of water consumption which is rarely available or reliable.

The design process has two well defined phases:

1. **Net tracing.** The plan of the water network is drawn on a map. There are practical limitations to the possible traces therefore normally pipes and other elements are placed on streets, along roads, etc. so they are more accessible and to avoid problems of property.

2. **Sizing the elements**, this is determining its size and properties. Pipe 32 will be 200 mm, Valve 3 will be a gate valve, the tank on the hill will be 40 m^3...In order to do this, it is worked out for the worst of the foreseeable cases, with the hypothesis that if it works in the most adverse case, it will work with no problem in the rest of the cases. In other words, if the pipe is 200 mm it can transport a discharge of 30 l/s of water, it will also be able to discharge 5 l/s.

Here is where EPANET comes in. By using this program, you will be able to determine which designs will behave adequately in the worst cases without needing very complicated calculations. Your task will be reduced to finding out which is the cheapest and most sensible among them.

3. **Repeat the sizing several times** trying different strategies that reach valid solutions.

4. **Benchmarking**. Make a comparative evaluation of the different designs to find the most economical one that is robust and practical. There is no need for a detailed budget, it is enough to establish costs per linear meter for the different diameters (pipe and excavations) and obtain an approximate total for each attempt.

CHAPTER 2

Thinking service, not infrastructure

Why do water projects?

The answer is pretty obvious: **to improve people's quality of life.**

Perhaps you expected a not very elaborated answer such as "so that people can drink" or a technical one "to achieve a peak flow of 28 l/s with a minimum pressure of 1.5 kgf/cm^2".

Stagnant waters from a leaking tap in Kabul, Afghanistan.

28 l/s may be a very noble objective however it could flood areas that cannot drain water with sufficient speed. The result is detrimental to the community to the point that, as

illustrated, even children put their hands on their heads in disbelief and ask themselves what on earth is happening!

In the effort to justify activity, deadlines or loss of funding, sometimes the real objective is missed: quality of life. We must not forget that whilst people's quality of life can be improved substantially, water projects **can end up being harmful to communities therefore** they should not be carried out carelessly or in a hurry, neither technically nor socially. Pay special attention to the **potential for conflict between communities** or social groups over access to the water source or the route of the pipes… Investigate potential discriminations.

Consequences of a poor access to water (and sanitation)

The idea is not new. Back in 1875 Joseph Chamberlain, the mayor of Birmingham, stated that the loss of working days, health and lives was costing the city £54,000 per year.

Some organizations have tried to evaluate the consequences of lack of access to water and sanitation. The figures are debatable, but the tendencies are clear. Some of them are shown below:

- **HEALTH** "Water related diseases are the cause of 80% of the deaths and illnesses in developing countries."[3]

- **ECONOMIC** "Benefits from Water and Sanitation programs refund between $3 and $34 per dollar invested."[4]

- **LABOUR** "Time saved in Tanzania meant a 10% increase in crop production."[5]

This last reference, *"Everyone's a winner? Economic evaluations of water projects"*, from WaterAid, is very interesting.

One of the greatest impacts is on people's health, which is very easy to measure because almost all medical installations keep statistics regardless of how basic they are. Some statistics to reflect on:

---

[3] *Secretary-General of United Nations Kofi Annan for the World Environment Day, 5th June 2003*

[4] *Hutton G. & Haller L. The Costs and Benefits of Water and Sanitation Improvements at the Global Level (OMS 2004).*

[5] *WaterAid (2004) 'Everyone's a Winner? Economic valuation of water projects' WaterAid, London.*

- "Diarrhea is the second cause of child mortality"[6].

- According to the WHO, **80% of illness in the world are related to access to water and sanitation**[7].

- According to the WHO, **there is no other measure with a greater impact on national development and public health as those related to water and sanitation**[8].

The human rights to water and sanitation

On 28 July 2010, the United Nations General Assembly declared that access to water and sanitation is a human right fundamental to the realization of all other human rights (resolution A/RES/64/292).

It is important to understand that this is not the typical first paragraph of a languid introduction with events and dates. The declaration has been a prodigious advance for humanity and has given a very important push to the sector. The declaration boosted the relevance on the international agenda, with many countries incorporating it into their national legislation. Imagine, **now we are human rights professionals!** Access to water and sanitation is a legal right for people, not just any commodity that you can buy, if you can afford it.

For this right to become a reality, it must meet a series of requirements that help us design interventions.

1. **Sufficient _and continuous_**: WHO recommends between 50 and 100 liters per day per person.

2. **Safe**, not only free of microorganisms, chemical agents, etc., but also in its location and access.

3. **Acceptable**, in terms of taste, smell and color, but also culturally and in a non-discriminatory way.

4. **Affordable**, no more than 5% of the family income.

---

[6] *Global burden of disease study 2020 .*

[7] *"Battling Waterborne Ills in a Sea of 950 Million", The Washington Post, 17 February 1997.*

[8] *WHO, factsheet 112 Water and sanitation.*

5. Physically accessible to all, including those with disabilities. WHO sets the maximum distance at 1000 m and the collection time at 30 minutes.

It is with this last criterion that I humbly think that perhaps we have lacked ambition. It is not realistic to think that 50 liters per person, much less 100 liters, are collected for a family of 6 people from a source 1 km away, not to mention for the elderly or people with disabilities. The reason is discussed below and explained in this video:

 https://youtu.be/Vq2c3P30FSA   (Base demand)

From infrastructure to service

The concrete objective of any water project is to provide a service using certain infrastructures. But too often, infrastructures end up getting all the attention.

The challenge is to change from a mindset **of installing infrastructure to one of <u>providing services.</u>**

If high chlorine concentrations, long distances to water points or excessive soap requirements due to water hardness make people prefer consuming polluted stagnant water, the system has failed to improve people's health, no matter how well its hydraulics work.

From infraestructure...

...to service.

User participation

Although in high-income countries such participation is not common, the level of services, the social and legal fabric, the existence of complaints mechanisms, and detailed planning do not make it as fundamental as it is in low-income countries.

Users must be able to **participate in the evaluation, planning, design, monitoring and maintenance of the systems**. Such participation is essential to avoid conflict, ensure the inclusion of all groups, adapt infrastructure to local culture and customs, address user concerns, develop a sense of ownership, and create appropriate maintenance attitudes. A significant percentage of the systems are subject to sabotage, vandalism and theft. The design process often involves negotiation and conflict resolution, and the **creation of mechanisms to collect complaints and objections from users.**

In water networks the simple thing is the design and calculation, it is the social part that has all the complexity and uncertainty! I wish there was a manual that structured it like this one for the social part. In the absence of such a manual, user participation is the best bet for the system to be viable.

Show your numbers!

Avoid also the opposite extreme, that is, that the social component is the only thing that counts and that capable technicians do not participate. If the calculation is easy, it is unforgivable that the benefits of a water network that has been socially worked in an impeccable way are lost because simply... it does not work. The funds, reputation, enthusiasm, social impact and trust that are lost every year simply because of this:

If you are a technician, show your numbers!
If you are a manager, outsource and show your calculations!
If you are a supervisor, show your calculations!
If you are a donor, ask for the calculations!

Even if you understand absolutely nothing and only file them! **It is essential to create a culture in which the calculations are taken for granted and to create the pressure for them to really exist.** Donors could play a major role here if they became aware of the waste of resources involved in eyeballed interventions.

In the cooperation cycle, donors do not ask for documents with calculations in the calls for proposals, very few NGOs have a technical department, the professionals who are hired do not usually have calculation skills because other skills are also required, and locally those skills are often not there either because the people who have them prefer to work in other industries or the institutions need to strengthen their capacity. In the end, they are often left undone, or the quality and level of detail do not make them very useful.

If you compare a project document in the context of cooperation to one coming from a high-income country, the difference is obvious. That there are no calculations is simply incomprehensible.

CHAPTER 3

Getting started with EPANET

EPANET as a tool for empowerment

A model is a construction that allows you to reproduce the behavior of a network in order to carry out tests and find solutions. In the case of EPANET, such construction is not a mock-up, but a mathematical representation of the relationships among its components. It is practical because it allows you run trials on "what would happen if…" saving lots of time, frustrations and money.

Other important advantages are:

- It prevents users from having to cope with our trial and error tests.
- It avoids social conflicts and getting into trouble with authorities.
- It improves the strength of the network because it avoids unnecessary installing and dismantling.

So, what is the reason for so few network models being built? The complexity of their calculations is overwhelming. Fortunately, it can be overcome easily by learning how to use the program which does all the calculations for you and allows you to concentrate on decision-making. Later on, I will explain why EPANET is used as a calculation program.

One of the biggest contributions of EPANET to development is making it possible for **people with little knowledge of fluid mechanics to make decisions regarding any given network**. At first, this may not seem very attractive, however, it is often difficult to find people in development work who are capable of making network calculations. The networks end up being taken on by people qualified in unrelated areas and locally, by people who have never had access to a solid technical training, especially in rural settings. Often, local people who have the technical skills to make such calculations don't have the motivation to live in remote places. They prioritize the education of their children, more challenging jobs, better living conditions, etc. and in any case, they aren't available in sufficient numbers.

The result is that interventions are undertaken anyway without a solid base and with questionable reasoning ending up with fairly bad results. In cases where networks are already built in populations with little resources, cannibalism starts. Parts of the network are dismantled in order to install them again in some other parts based on impulses and hunches with little basis.

In my opinion, a more pragmatic approach should take priority. Often Development Agencies believe that NGO's will have the ability to calculate networks. NGO's then assume that such ability can be found locally, which is not the case, especially in the bush. The result is that nobody has the capacity and the network goes undesigned…do you not think it would be more appropriate to supply people who are directly in charge of the networks with the necessary tools to do so?

The primary objective of this manual is to facilitate this empowerment. The first edition was also used by many engineering students around the world. If you are one of them, welcome aboard, I hope you find it useful.

Introducing EPANET

EPANET is an easy to use calculation program distributed by the U.S. Environmental Protection Agency (EPA), which has a very visual interface and intuitive operation. It can be downloaded for free from the US Environmental Protection Agency website. As you can see in the image below, it is not that intimidating.

In fact, it is very easy to use, which has contributed to its popularity. I have learnt from experience that people without prior knowledge of EPANET can learn the basics in about 8 hours and can learn how to use it in about 30 hours.

EPANET is currently in version 2.2, with version 3.0 under development. The development of the program, which stagnated after version 2.00.12 for some time, is now done as open source by a handful of volunteers. The changes mainly affect the calculation engine and not the system functionalities (such as implementing an undo button or integration with other programs such as AUTOCAD that are made with third party applications).

In a way, from the user's point of view, Epanet is frozen in time. Its development was stopped in court by the very companies that use its calculation engine, claiming unfair competition from a public entity. That can give you the impression that it is outdated and its use is more academic than anything else. Nothing could be further from the truth. Epanet is the default industry standard, used in third-party software and is used to design billions of dollars in water infrastructure investments.

EPANET has been translated into English, Spanish, French, Portuguese, Russian and Korean. Most of these versions are previous to 2.2 but using them is not a problem. I have used the earlier versions for 20 years to design systems for millions of people.

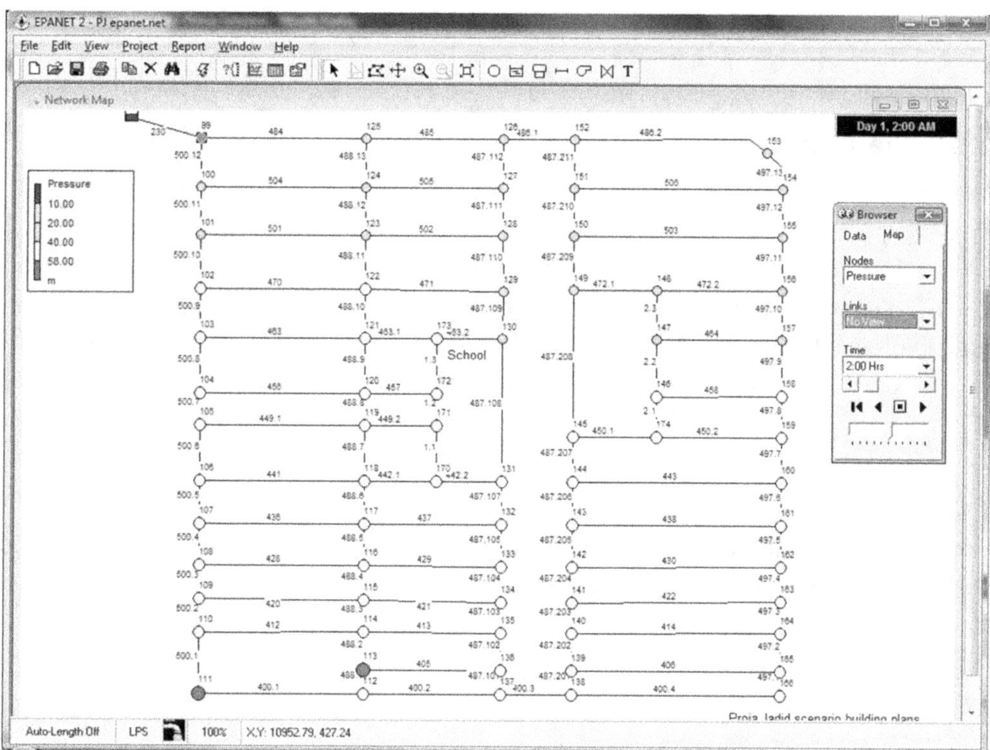

Some Practical Information

Downloads

English, with the latest up to date version:
www.epa.gov/water-research/epanet

Spanish v2.00.12:
www.iiama.upv.es/iiama/es/transferencia/software/epanet-esp

French v2.00.10: The download of the program itself is somewhat hidden, look in the "Supplementary resource" box.
https://www.researchgate.net/publication/330525216_EPANET_20_en_Francaise_MANUEL_DE_L'UTILISATEUR_Version_20010

Portuguese v2.00.12:
www.dec.uc.pt/~WaterNetGen/epanet_por.php?DownLoadEpanetPor=Nothing

Russian v2.00.12:
www.epanet.com.ua/download

YouTube channel

 tiny.cc/arnalich

In my YouTube channel there are videos about design, how to model boreholes, springs and pressure rupture tanks with EPANET or about the practical use of a hydraulic slope of 5 m/km to optimize networks in which you don't even know where to start.

Exercises

To our knowledge and at the time of writing there is only one manual with exercises which we also prepared, *EPANET and Development: A progressive 44 exercise workbook* available at: www.arnalich.com/en/books.html:

Add-ons

Epanet has many add-ons and tools developed by third parties, for example, several for working with AutoCAD or GIS, or those developed by Oscar Vegas Niño. Look for them on the internet.

If you can't see the help files

Some versions of Windows, from Vista to the first versions of Windows 10, do not manage to read the help files and the tutorial, which are frankly useful. If you are in that case, look on the Internet for an updated solution. For example:

www.water-simulation.com/wsp/2015/10/01/how-to-open-epanets-help-file-in-windows-10/

What EPANET can and cannot do

Good news! EPANET can do most of the calculations you may need for your project, and those that EPANET cannot do are actually pretty easy to do on a piece of paper. You will use EPANET mainly for:

- Determining what pipes with which diameters should be used.
- Determining what improvements and /or extensions the network needs.
- Determining where to install the tanks, valves and pumps.
- Studying chlorine's behavior and the necessity to establish secondary chlorination points.

Although you can also use EPANET to make the following calculations, in my opinion, it is safer and much quicker and less prone to error to make them by hand:

- Dimensioning tanks.
- Pump sizing, except for complicated systems.
- Estimation of energy consumption.

EPANET's benefits are described in detail in the EPANET User's Manual.

What EPANET CANNOT do

EPANET assumes the quasi-equilibrium condition, that is, it assumes there are no abrupt changes in the network. This is true up to a certain point. Long pipes (many kilometers long), are very resistant to change, and users do not behave like a sardine shoal, opening and closing taps in unison. However, it leaves out some real and quick phenomena such as pipe bursting, water hammer, sudden shutting of non-return valves, starting a pump or stopping it, etc.

Some of the things EPANET CANNOT do:

1. Calculate water hammer.
2. Simulate pipe bursting, it only simulates leaks flow and its effects on pressure.
3. Model the real behavior of non-return valves.
4. Evaluate the consequences of the presence of air inside the network.

To sum up, you can't model sudden and fast-occurring changes.

EPANET´s main objects

EPANET recognizes 6 types of objects that are found in networks and it is vital to know them as the network is modeled on them.

They are shown and explained in the following paragraphs:

○ **Junctions**. A junction is a point at a certain height, where water can leave the network. This outlet is created by assigning it a demand. When a negative demand is assigned, it is automatically turned into an inlet. A borehole can also be shown as a junction where the height represents the water level inside. In junctions, demand is known and pressure is unknown.

▱ **Reservoirs**. The reservoir works as a drain or as a water source. It is a good idea to use one when working with EPANET, to avoid error warnings. The volume remains constant regardless of water input or output because of its huge size in comparison to the system. In order to picture what they could be like, think of rivers, lakes, aquifers…

▣ **Tanks** have a limited capacity to store water and the water level increases or decreases as they fill or empty.

— **Pipes** convey the water from one part of the system to another. EPANET assumes that pipes are always full. Furthermore, it assumes that by using their properties they are capable of being opened or closed, and limiting the flow to one direction therefore it is not necessary to add check-valves to the model.

⌒ **Pumps**. If there is one thing we should be scared of when working with EPANET, it's the pumps. It is wise to avoid them as much as possible (by creating equivalent gravity-flow systems), as they cause too many unexpected headaches that may end up testing your nerves. Pumps impart energy to the water, in other words, they lift it.

⋈ **Valves** as understood by EPANET are elements best avoided in developing contexts due to their high cost and difficulty in obtaining spares. Remember that non-return valves (to prevent backflow) and shut off valves are already included in the model as a property of the pipe. There are several types of valves:

- Pressure reducing valve (PRV). Limits the pressure to a value.
- Pressure sustaining valve (PSV). Maintains the pressure upstream to a given value.
- Pressure breaker valve (PBV). Forces a pressure loss across the valve.
- Flow control valve (FCV). Limits the flow to a maximum value.

- Throttle control valve (TCV). Simulates a partially closed valve.
- General purpose valve (GPV): the user designs its behavior.

Avoiding wasting time with EPANET

EPANET is used to save time and effort. Try to avoid:

1. **Drawing networks too precisely**. When you work with the automatic length option switched off you are drawing a sketch of the network to which you will introduce lengths separately. EPANET does not care if your drawing is accurate or not and will give you the same results for these two networks providing you have given the same length for each pipe:

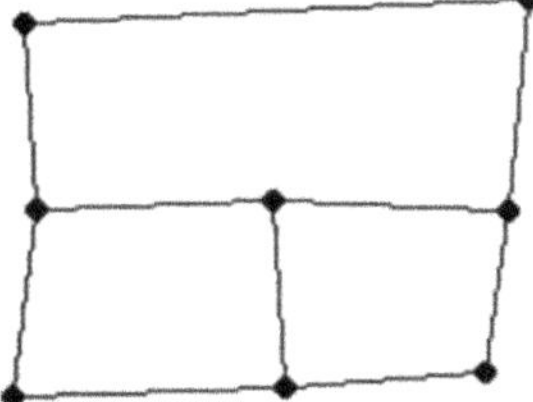 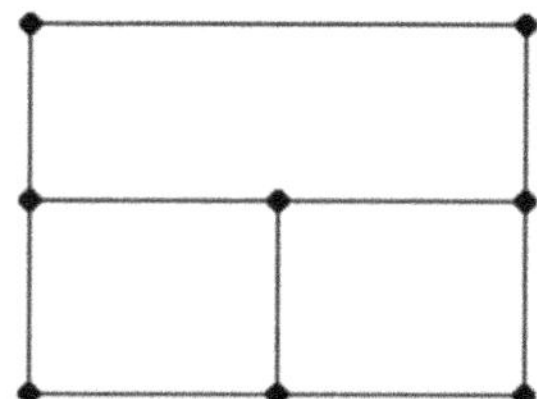

2. **Labeling pipes and junctions** with logical names. During the process, EPANET assigns one number to each object: pipe 1, junction 63 and so on. To attempt to make these numbers follow certain logic is not a good idea as during the design process you will erase and add lots of pipes. Updating the labels would be a very time-consuming task. With the model already finished, you can use Oscar Vegas Niño's Rename-IDs tool to rename them as you like.

3. **Destroying the template model**. Once you have introduced all the topographic data, pumps, etc., save it and work on a copy, especially for existing networks as after hours of fruitless work on a model it is difficult and time consuming to revert back to the original.

4. **Making changes to the model of an existing system without writing them down**. Congratulations! You have managed to get an optimized design, with an adjusted budget, buuut….what did you change on the 267 pipes and 198 junctions?! Write down any change. Write something like this:

   Pipe 58, changed from 75 to 125 mm
   Pipe 63 new
   Pipe 113 deleted…

For example:

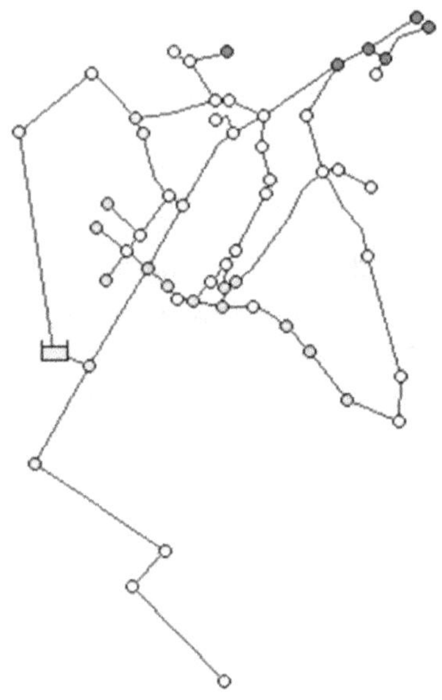

You should:

1. **Use default options** for the lengths, diameters, demands, etc. that are most common in your system. Later on you will see how to do it.

2. **Save versions of the file** as you complete the steps. EPANET does not allow you to undo changes automatically. The possibility of going back to the starting point after trying to model something which has failed will save you a lot of time. Saving the networks as "Network v1.2 stable.net", "Network v1.3 pumping station.net", etc. allows you to go back to the previous point quickly.

3. **Model simultaneously in single-period and extender period** by using the option starting time of the analysis. This will be covered in the section "One way of working" in chapter 7.

If you feel like you are full of energy at the beginning of your project and you want to be so meticulous that you decide to ignore some of the recommendations above, you should read the section "Overcoming the Post-Assembly Laziness Syndrome" in chapter 7. After reading it, you may reconsider saving some energy for later on.

Setting the units

On September 23, 1999, the Mars Climate Orbiter crashed on the surface of Mars because the ground and probe computers were working in different units. To avoid similar embarrassments, **make sure EPANET, your team, the suppliers and the contractors are using the same units.** The first thing you must do when you start the program is to configure EPANET to make sure it will work with the right units.

In order to do this, click on Project and, in the drop-down menu, select Defaults. The rest of these instructions are given in the manual by means of routes. The route for this action would be: *>Project/Defaults.*

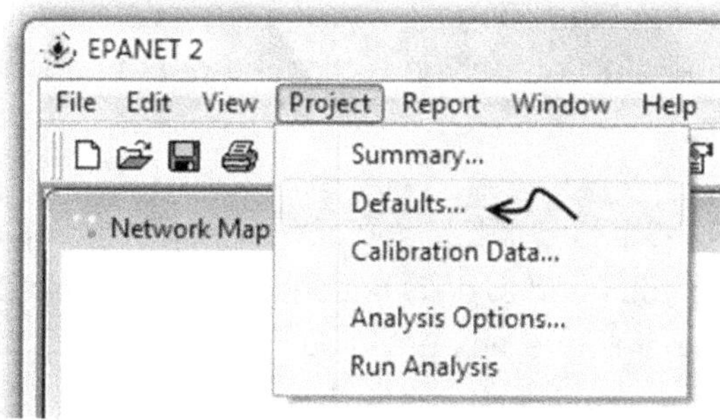

In the menu that appears, click on the Hydraulics tab and make sure that you put LPS. Choosing this option implies that the following units are used:

- Flow Units: liters/second
- Pressure: MWC [9]
- Diameter: millimeters
- Length: meters
- Height: meters
- Dimensions: meters

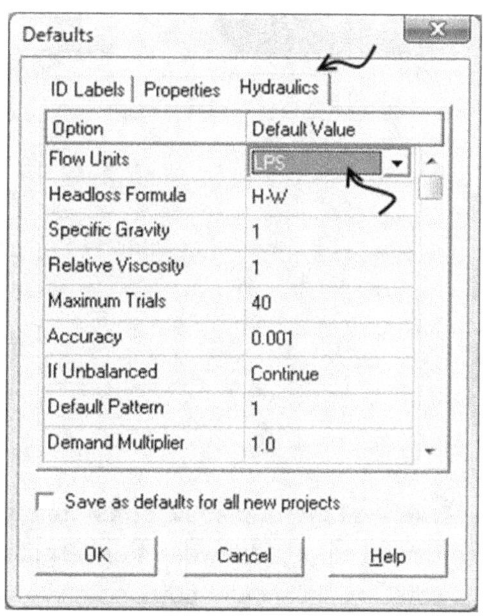

---

[9] *Meter Water Column, is a convenient way to measure water pressure and relate it to the topography. 10 meters of water column is a bar, another frequently used pressure unit.*

Basic tools

The object of this section is to highlight some important tools and how to make the most of them.

- ➤ **Editing by Regions**. If you repeatedly have to change the pipes roughness from 140 to 120, it is not necessary to do it pipe by pipe:

    1. Follow the route: >*Edit/ Select Region*. You will see that the cursor becomes a cross.

    2. Select the objects you want to change. Surround the selection area by clicking around it. To close the selection polygon, click the right button of the Mouse. To start again click Esc.

    3. Once it is closed, click >*Edit/Group Edit* and have a look at the menu that shows up. Now, the only thing you need to do is to construct a sentence by ticking and selecting options in the dialogue box until you come up with a sentence such as *For all pipes within the outlined area with roughness equal to 140 replace roughness 120* as in the image below:

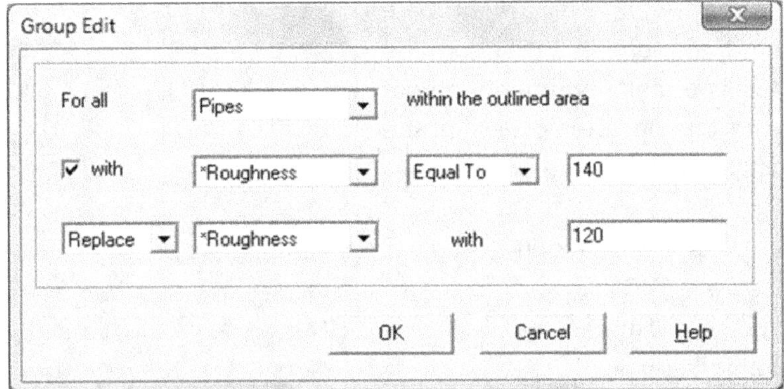

- ➤ **Make the legend´s scale consistent with design criteria**. If you have decided that pressure must be between 1 and 3 bars (10 to 30 meters), you can modify the legend so that the first color shows negative pressures, the second one shows the bottom limit, the third one an intermediate level, and the fourth one the maximum pressure level. This way you can detect values that are out of order with just a quick look.

In the image on the right, there are zones with negative pressure (dark blue), insufficient pressure (sky-blue) and excessive pressure (red); apologies if you are reading the printed B/W copy.

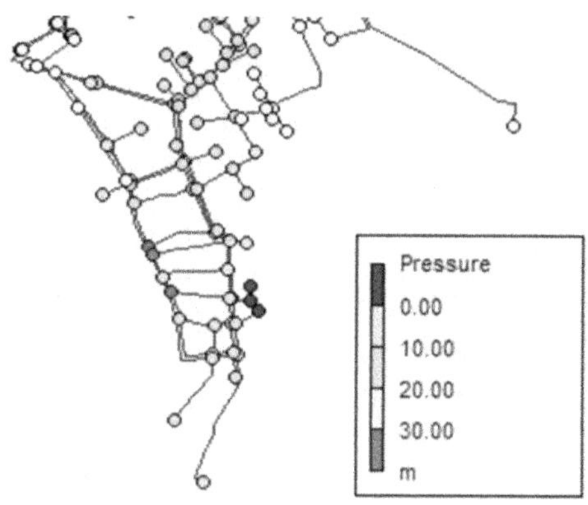

The procedure is:

1. Click the right button on the legend. If you click it with the right button, it disappears. To display the legend again, go to View in the toolbar, click on *>Legends/Junctions* (or *>Legend/Pipe* depending on the parameter you are referring to).

2. Write the values of your choice in the textboxes and click OK.

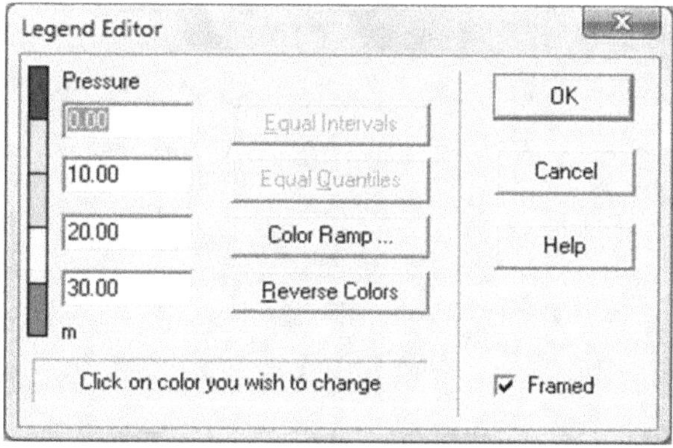

➤ **Use the Help**. EPANET's Help is surprisingly good and contains a lot of useful, quickly accessible information:

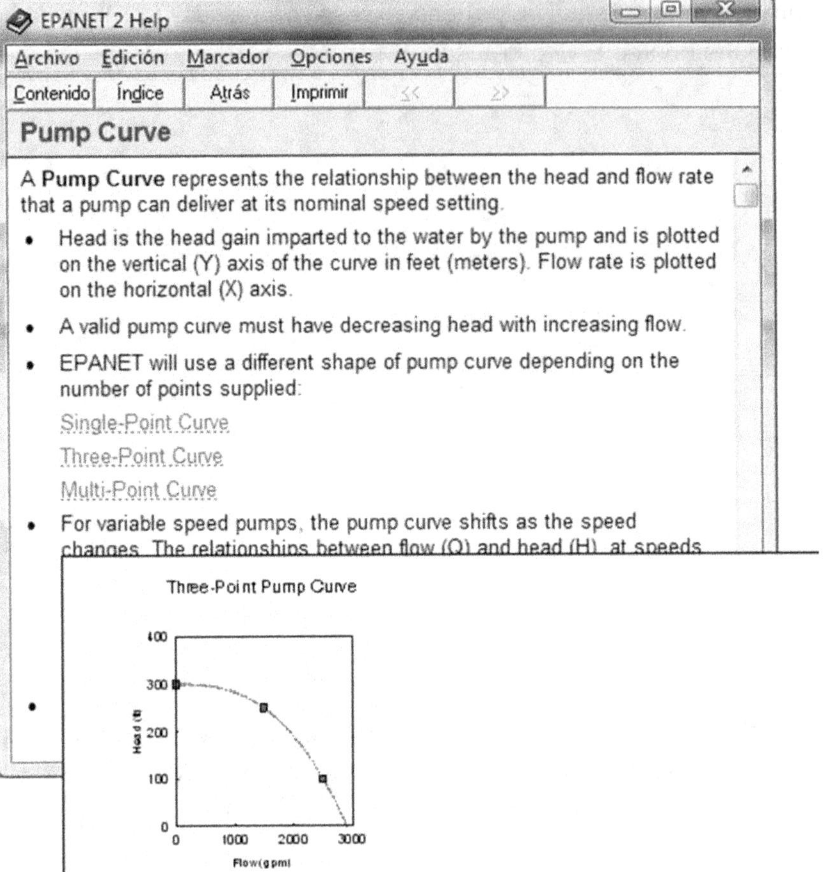

Pump Curve

A **Pump Curve** represents the relationship between the head and flow rate that a pump can deliver at its nominal speed setting.

- Head is the head gain imparted to the water by the pump and is plotted on the vertical (Y) axis of the curve in feet (meters). Flow rate is plotted on the horizontal (X) axis.

- A valid pump curve must have decreasing head with increasing flow.

- EPANET will use a different shape of pump curve depending on the number of points supplied:

  Single-Point Curve

  Three-Point Curve

  Multi-Point Curve

- For variable speed pumps, the pump curve shifts as the speed changes. The relationships between flow (Q) and head (H) at speeds

A three-point pump curve is defined by three operating points
- Low Flow (flow and head at low or zero flow condition)
- Design Flow (flow and head at desired operating point)
- Maximum Flow (flow and head at maximum flow).
EPANET tries to fit a continuous function through the three points to define the entire pump curve

CHAPTER 4

Assembling the model

Introduction

In this chapter we will build a model. We start by drawing, and then add the data. The estimation and allocation of the demand is covered in the next chapter, "loading the model."

Before you start, please take an hour to follow the tutorial which comes with the program to familiarize yourself with the program. By following this tutorial you will be ready to continue with the rest of the manual. It can be found in: *>Help/tutorial*

If you are using Windows Vista or later, read the section 'Lost help files in windows in Windows Vista and later' in chapter 2. The tutorial is also in the manual.

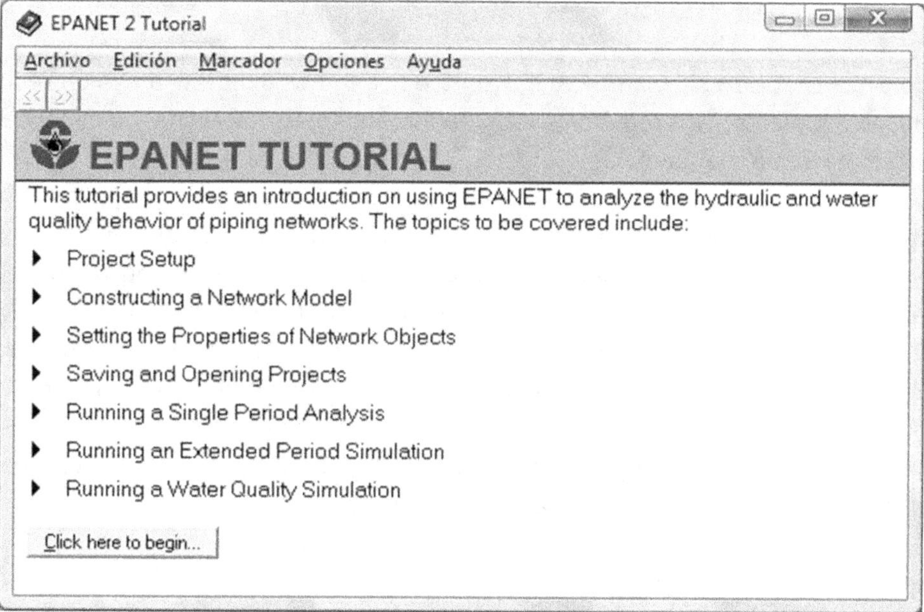

Finally, **make sure that you have configured EPANET´s default values and units**, as seen in the previous chapter.

Sticking to units and avoiding errors

To work with EPANET you need to do some very simple calculations by hand. These although very simple can prove tricky, like double negatives or finding out the number of days between 2 dates.

If you are disciplined and carry the units, you will be able to detect many of these mistakes before they affect your emotional stability. Units are your canary in the coal mine. Take a look, for example, at these two calculations for the same unit conversion:

$$14m^3/h = 14\frac{m^3}{h} * \frac{1m^3}{1000l} * \frac{3600s}{1h} = 50.4\frac{m^3 * m^3 * l}{h * h * s} = 50.4 lm^6/h^2 s$$

l*m^6/ h^2*s?! If, like me, you don't know this unit for flow, you know something went wrong.

$$14m^3/h = 14\frac{m^3}{h} * \frac{1000l}{1m^3} * \frac{1h}{3600s} = 3.88\frac{m^3 * l * h}{h * m^3 * s} = 3.88 l/s$$

Adding cartography

There are 2 ways of working with EPANET. In the fist you would draw a sketch of the net and introducing the data. This requires a topographic survey to determine the distances and elevations. In a word it's laborious.

The second way which is the main focus of this section is easier and more accurate. Basically, you download a map or satellite image and set it up as a backdrop in the program. That image is calibrated by entering the actual horizontal and vertical dimensions, so that EPANET can establish a relationship between pixels and actual distances, for example, 500 px/km. With that relation EPANET will estimate the lengths of the pipes you draw on top.

You will most likely start with images from Google Earth, but let's start with printed maps, to make the explanation about the coordinates more visual.

Finding out map dimensions

Maps usually have this distance already measured in the form of coordinates.

    a)   Recent maps will use UTM and its margins read something similar to this:

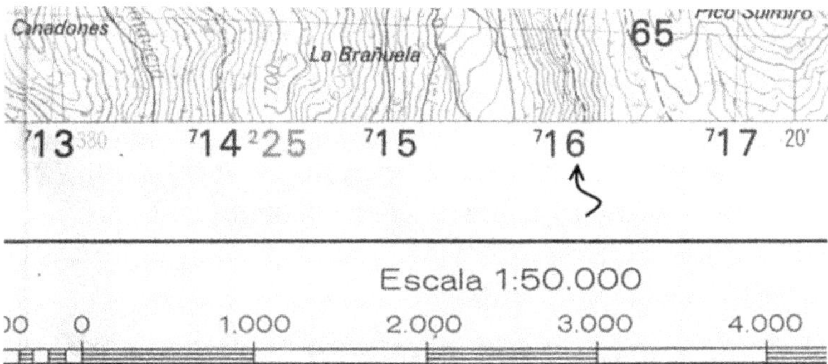

Because the scale is given, it is very easy to deduce the dimensions. If this was not the case, look at the numbers ⁷13, ⁷14, etc. These represent the horizontal UTM coordinate. In a GPS, ⁷13 will appear like this:

32S **713**000 8033400

You can ignore the first group, 32S. The second group is the horizontal coordinate measured in meters from a point, and the last one is the vertical. In order to work out the distance between ⁷13 and ⁷14, you can subtract using this following method: point ⁷13 is 713,000 meters from the reference point (which we are not interested in) and ⁷14 is 714,000 meters away. Therefore, the distance between them is 714,000 - 713,000 = 1000 m.

Let's look at an example. The framed area of the map on the following page is where we plan to do a project and we need to calculate the dimensions.

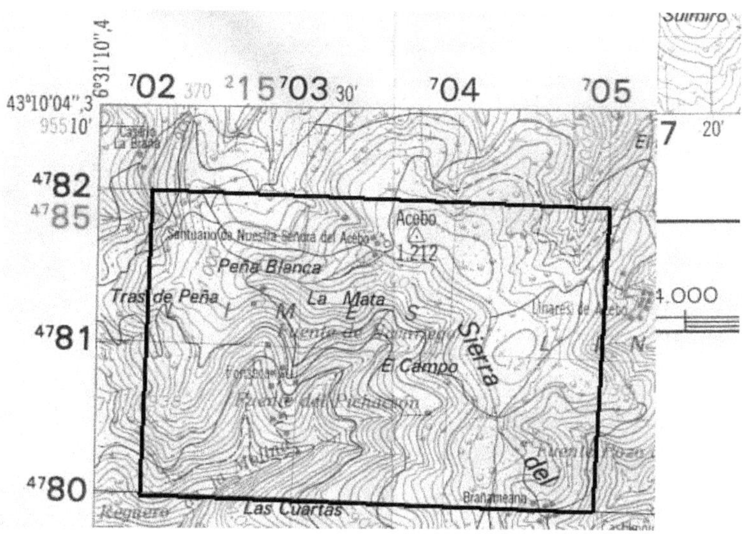

Horizontal dimension 705,000 - 702,000 = 3000 meters
Vertical dimension 4,782,000 - 4,780,000 = 2000 meters

One big limitation of EPANET is that it only accepts border dimensions therefore the map needs to be trimmed to show only the black rectangle of the area to be worked on.

b)   If the map is given in longitude/latitude degrees, you can use a coordinate converter which can be found on the internet, and proceed the same way.

Establishing dimensions of images and maps with no references

You need a GPS set for UTM coordinates or use a satellite image software like Google Earth to find out the coordinates. Do not use latitude and longitude, as they are only practical in open areas such as sea or air navigation. As seen before, your UTM coordinates will be something like this: 32 S 486000 8033400. The first coordinate, 486000, is the horizontal in meters, the second one, 8033400, is the vertical also in meters and the group 32 S has no use at this scale.

If for example you need to calibrate the following image simply locate 2 points which are easily identifiable on the map and in reality. In this instance the gas station and mosque are ideal as they are located at each corner of the map and therefore will cover the complete area.

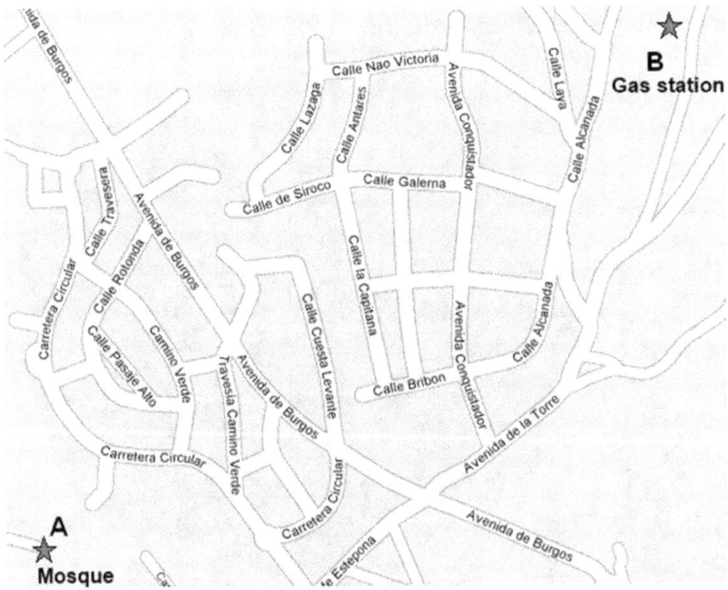

Once the points are chosen go and take their UTM coordinates. The subtraction between the two coordinates is the dimension of your future image:

Gas station (upper right corner)   32 S 486000 8033400
Mosque (lower left corner)   32 S 484000 8032000

2000   1400 meters

Remember, it is important when using EPANET that the points are in the corners of the rectangle as the image will be cut to coincide with these points.

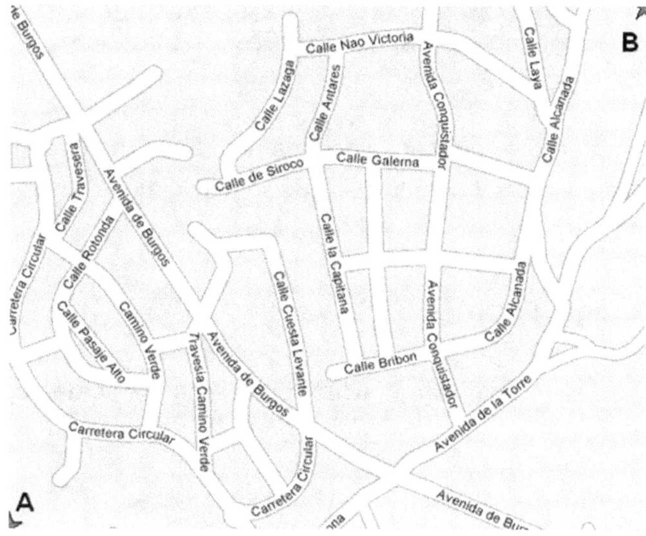

In order to load your image to EPANET go to >*View/Backdrop/Load* and search until you find it. Some versions of EPANET work only with .bmp and some less common formats. To change the image to .bmp use the option 'Save as' (as in all image programs, including Paint). Once you have the backdrop image, set its dimensions. Click *View/Dimensions* and fill in the dimension dialog box like this:

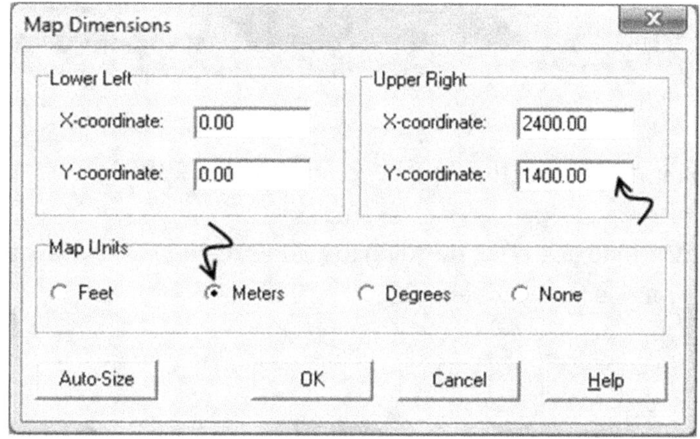

Alternatively, you can insert the whole coordinates (i.e. 486000 for X coordinate) and let EPANET take care of the subtraction. Coordinates will appear on the lower right corner of the program when you move the pointer.

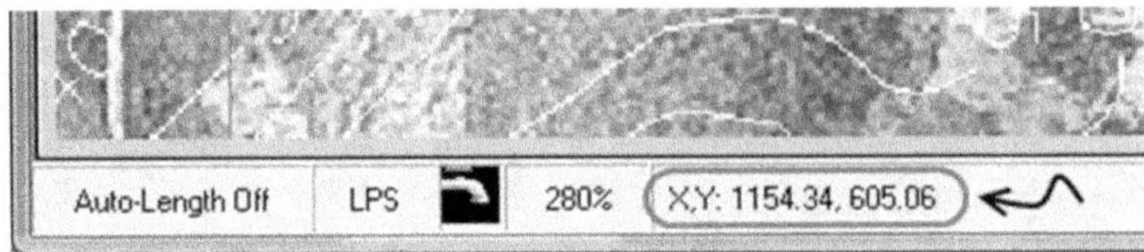

To confirm that you have entered it correctly, place the pointer on the upper right corner of the image. Allowing for a few meters it should coincide with its dimensions.

Using images from Google Earth

From Google Earth you can save an image of what you see on the screen by following the >*File /Save /Save Image...*:

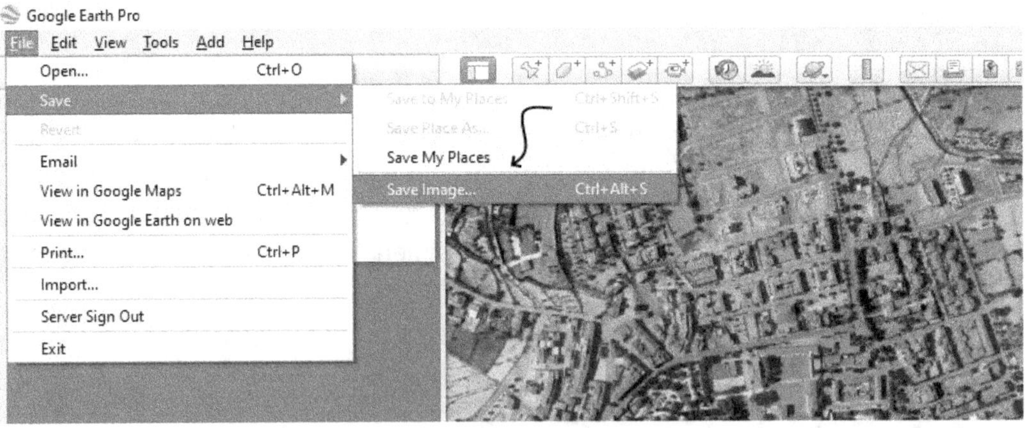

In order to find out the dimensions of the image, you can do the following:

1. Set up Google Earth to use UTM by going to >*Tools /Options /3D View* and checking the *Universal Transverse Mercator*.

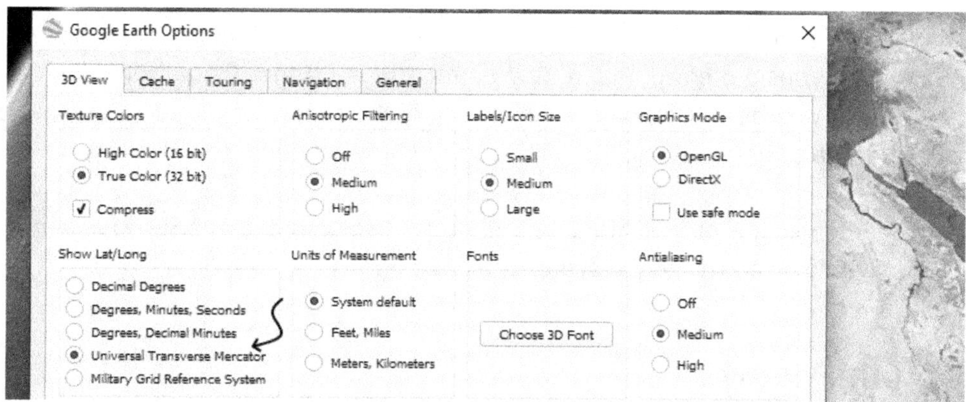

2. Add a pin to the lower left and upper right corners as done a few pages back. With this you can mark the points where you should cut the image, at the tip of the pins, and get the coordinates at the same time.

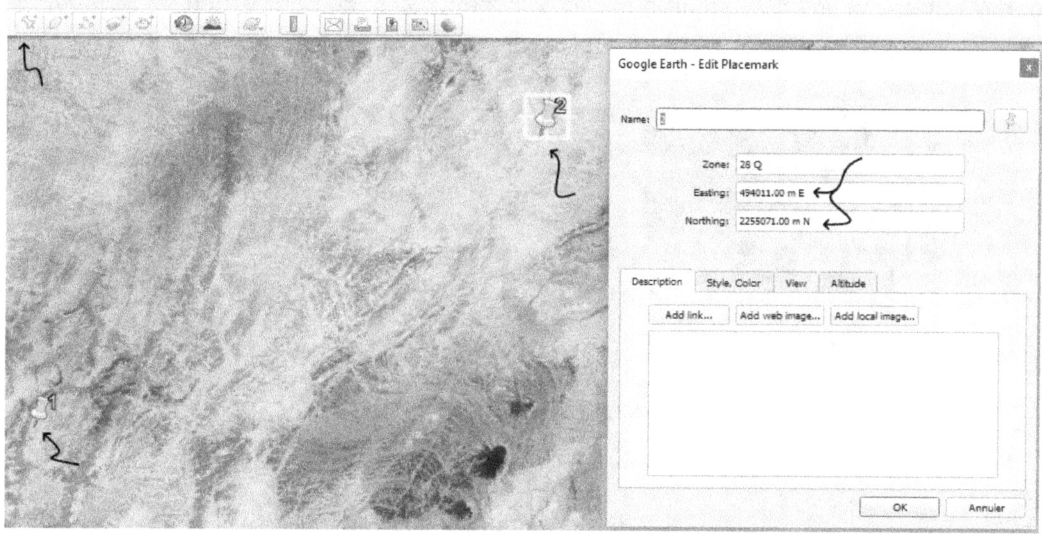

3. The rest of the process will be the same as shown earlier.

Some practical matters

a) You are already with the referenced image. You start drawing the network and... Oops! You can hardly see anything! It happens mostly over vegetation. In >*Browser /Map* select to show *Initial Quality* and *Wall Coefficient,* for example, and modify the scale if necessary, so that what you draw is seen in red or another high contrast color.

b) If your version requires you to use the .bmp format, the images will have a large size. Use your preferred image size bearing in mind that the larger the size, the slower the operation. Personally, I think 2 or 3 Mb images (or even 10 Mb depending on the area to be covered), are a fair compromise between detail and agility.

c) In order to use this process trouble free do not use distorted images. Use images in which vertical and horizontal scales are the same.

Drawing the network

Over a calibrated backdrop

When you have loaded a backdrop and set its dimensions you can draw directly on top of it. Check frequently that the auto-length mode is activated when you draw. It has an insistent tendency to become inactivated. In order to activate it, click the right button on *Auto-length Off*:

It is inevitable however that at some stage you will find yourself drawing with the mode inactivated. To detect recently added pipes where auto-length was not used, check with the default pipe length you have, normally 1000 m. You find this value in *>Project/defaults*, in the *Properties* tab. Then, make a 'query' with the icon:

Once you have filled in the fields *Find links with*, *Length*, *Equal to* and the value, click submit. The items that meet the conditions will appear in red as shown in the image. The arrows have been added in order to highlight the links, but they will not be shown in EPANET. It would be very unlikely that the pipe you have drawn on auto-length is exactly 1000 m therefore you can conclude that these pipes have not received the real value and they are inheriting the value by default.

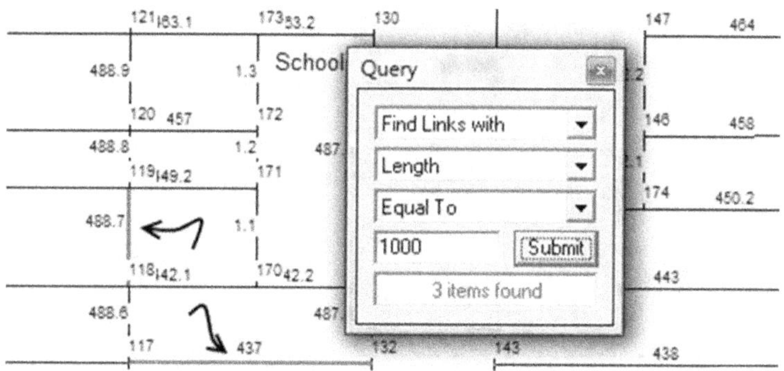

An important point to remember when calculating distances is that EPANET will not take into account the altitude of junctions but treats it as a flat surface. In the image below EPANET would calculate the length of the pipes A, B and C as 500 m, although more pipe would be needed to reach the point 100 m higher than the horizontal. The overall pipe requirement will not change dramatically but **remember to order an additional 5-10%** (see chapter 8). This way, you will not find yourself needing 25 extra meters and having to wait 6 months for the next order to arrive.

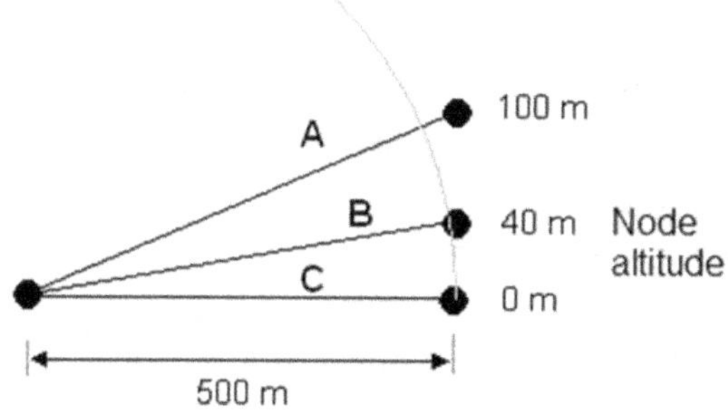

With a topographic survey

Unless your network is located in a very flat area, you will need to measure the relative height of the junctions. Even when you have cartography, it is convenient to predict which high points in the pipeline will need air released and which low points will need sediment removed. A topographic survey looks like this:

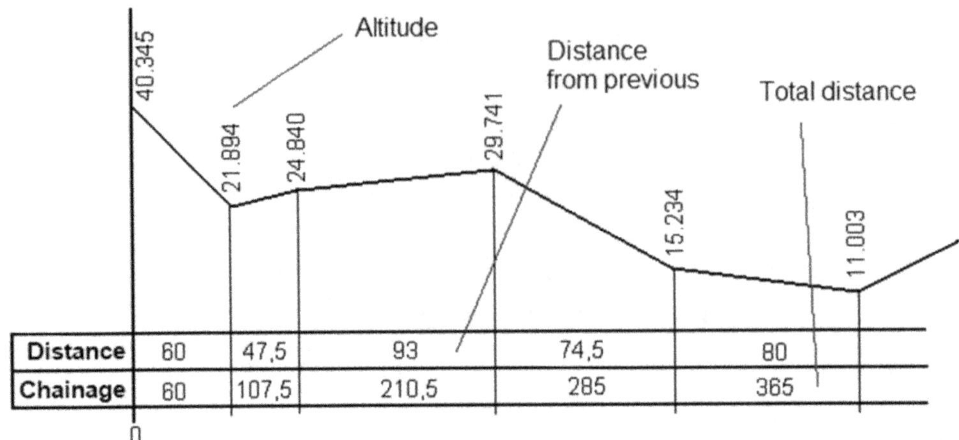

It basically shows the height of a point and the length of the pipe between two points. For example, from the first point to the last one, there would be a pipe of 365 m (365 - 0) and the height difference is 29.342 meters (40.345 -11.003 m).

Some people suggest (I did myself at the beginning of this book) that you can save a lot of time by introducing the most common length of the pipe by default in repetitive networks (remember: >*Project/Defaults*).

If you do it this way, be careful because when you draw a sketch that does not follow a scale, it will be difficult to trace pipes with unrecorded data (see previous chapter). For example, in the image below, the two pipes indicated by arrows seem to be similar lengths; however, the first was 90 m whilst the second was only 40 m. Once you have put in an existing default length how can you check to see if you have forgotten to input any data?

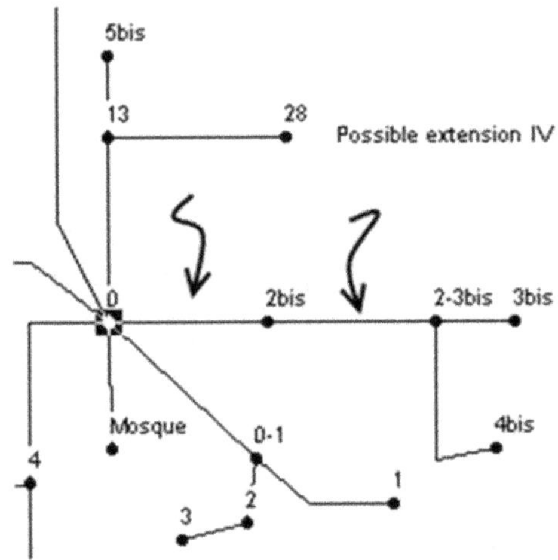

Importing maps from AutoCAD

In some cases, the network may be digitalized in AutoCAD. There are some programs that interpret the information contained in AutoCAD and transform it so that it can be understood by EPANET. One of them is dxf2epa, developed by Lewis Rossman, creator of EPANET. A newer software and probably a better choice is EpaCad, available at: www.epacad.com

It is important to pay attention to these two things:

1. If you have not used accurate tools with AutoCAD (such as Snap) lines which appear to be connected but in reality are not will be disconnected in EPANET.

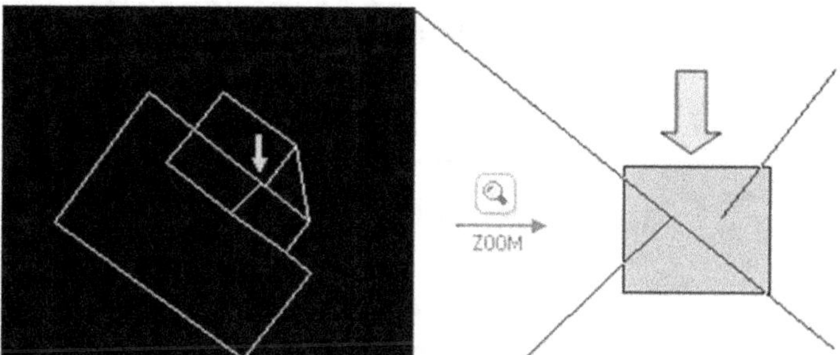
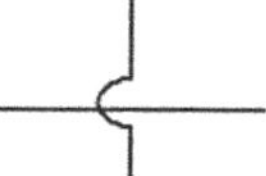

2. Crossed lines are interpreted automatically as a junction. If we have crossed pipes which are not connected, we need to disconnect them by representing them as it shown on the right.

Downloading maps from Google Earth

With a pro version of Google Earth, which is free for nonprofits, you can obtain satellite images by following Archive/ Save/ Save image. Ask a search engine for other alternatives, if you cannot get the PRO version. Frequently a simple screen capture will do the job.

Including details

Keeping the scale within the picture can cause congestion as with the pumping station shown in the image below. The same applies if you create a small system in the same network such as schools, hospitals etc.

The easiest way to deal with this is to deactivate the automatic length, draw this part of the network enlarged then assign the real lengths afterwards. In the image, the area inside the circle corresponds to a pumping station and although in reality it should only occupy a few meters, to legibly model all its components, it has been enlarged to avoid having an unintelligible cluster of points and pumps.

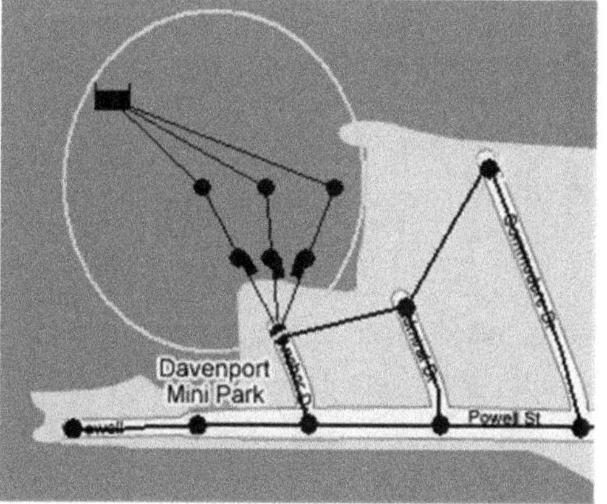

Entering data for junctions

Now we have drawn the network, it is time to go over the data we will need. As you will know from the tutorial, when double-clicking on an object a dialog box opens and shows its properties. In this box, you can read some things and modify others but cannot modify all of them.

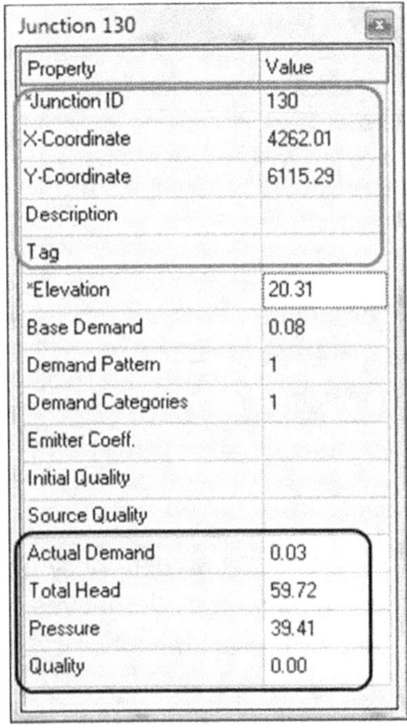

This is the complete dialog box for a junction, with all its properties.

Those which are shown inside the red square (above), can be modified, but will acquire an automatic value when the object is drawn. These properties are self-explanatory.

In black (underneath), is calculated data which is read-only. Any required data comes with an asterisk which in the case of a junction would be elevation and name.

Focus on the parameters in the center:

1. Elevation
2. Base demand
3. Demand Pattern
4. Demand Categories
5. Emitter coefficient
6. Initial quality
7. Source quality

These parameters will be covered in the following sections. The main part, the demand (points 2, 3 and 4), are covered in the following chapter. Don't be intimidated, it is actually pretty simple.

Elevation

The elevation is simply the height of a given point. Because we work with relative height, it makes no difference if the heights refer to sea level or to a stone where you may have forgotten your hat. The reference point which is used to compare elevation points is called the datum. Normally, for example with topographic surveys, this will be the base of a tank or a borehole.

Imagine you do not know the height above sea level for the following three points in a network (they have been put in a column as a reference). If you decide that the datum is going to be the borehole and you assign it an elevation of 0 meters, you will have:

Object	Relative height	(Height above sea level)
Datum Borehole	0 meters	500 m
Tank	67 meters	567 m
Spring 1	22 meters	522 m

If you used the tank as the datum, you would have:

Object	Relative height	(Height above sea level)
Datum Tank	0 meters	567 m
Borehole	- 67 meters	500 m
Spring 1	- 45 meters	522 m

The most accurate way to determine heights is by doing a topographic survey. **Never use a barometer, Google Earth or a conventional GPS (<8000 USD) in order to determine heights**. These devices have an error of ± 10 m at best, which is the whole pressure range of a well-balanced system. You may use them for preliminary studies but not for the final design.

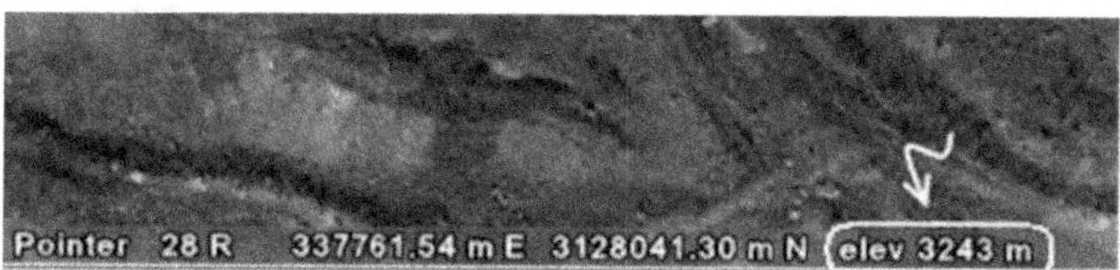

The same applies for maps with contours and free digital elevation models (DEM) on the internet. For example, with the free program Google Earth, navigate until you can match the UTM coordinate of a certain point in the navigator and read its height.

If you use these, be cautious and check the accuracy in your area. In some places and moments, you may get surprises, like the sea level is at 17 m:

It is human nature to forget one or more of the points. Oversights will have a value of 0. Repeat the search in 'query':

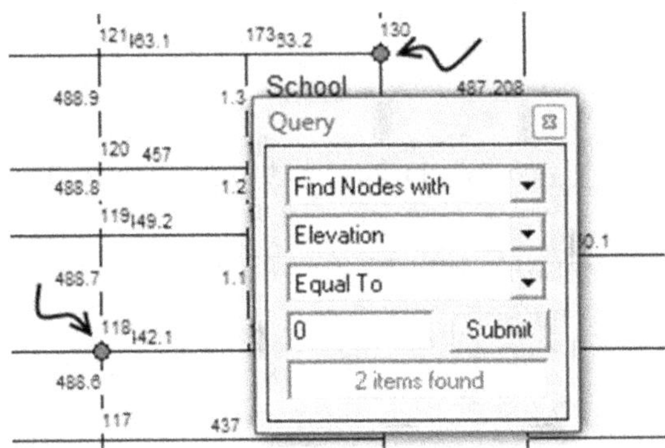

Watch out with "tricky" elevations

Determining which elevation you should use may be tricky. Read the following paragraph carefully, because sometimes it is not just as simple as determining the elevation of a certain point. In a model, not all the connections to the houses are represented, nor all the points through which a pipe passes. You must make sure that there is enough pressure and that the flow is not interrupted at any time.

A house at point C may not receive water.

Remember to put "pressure gauge" nodes to read the pressure at the highest points of the network.

When determining elevations, it is not necessary to determine the exact elevation of a pipe. Use ground level as an approximation, the only error being the pipe's depth which is less than a meter.

Following the same reasoning, if the pipe has a point, A, which is higher than the water pressure line, it will not be able to reach point B.

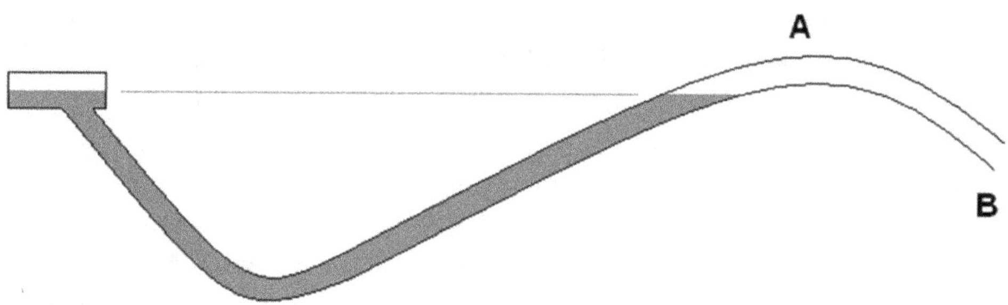

In order to prevent this happening, although problematic, place a junction with no demand in the highest point of the pipe. Once the water starts moving, the hydraulic gradient will not be horizontal any more. It is more difficult to evaluate the pressure in point A if there is no junction. Except for some particular and exceptional cases, the minimum pressure of these points must be 10 meters in order to avoid pressure problems.

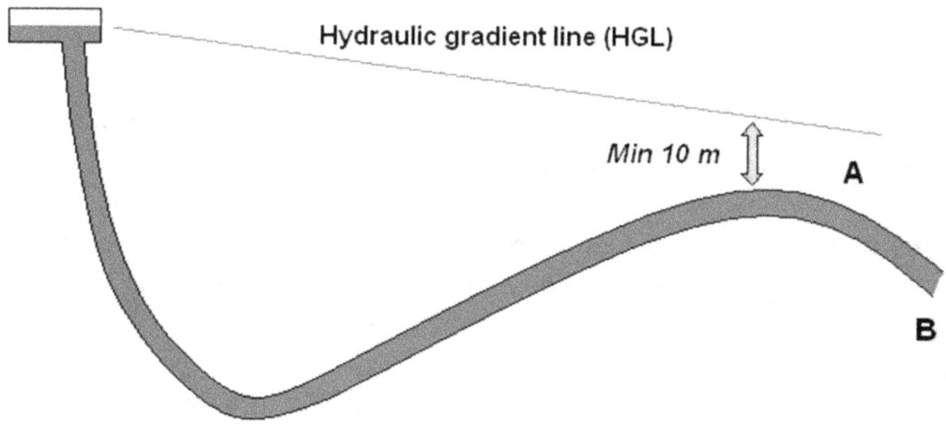

What are these exceptional cases?

In cases where the pipe cannot be re-routed and would need pumps or elevated tanks in order to meet the requirement. In these circumstances, if the pressure in point A is a least 5 meters, it is not worth adding complexity to the network by adding a pumping station or an elevated tank.

Emitter Coefficient

They are used to simulate free or pressure-dependent flows by avoiding specifying a specific demand, such as leakage, irrigation or the free discharge of a pipeline. For the latter case, for example, a very high coefficient type 9999999 is set.

Initial Quality

This is usually left blank. It would be used when modeling if you were required to start with a value for water quality, for example, 0.6 ppm of chlorine. Its main purpose is to avoid long waits before the chlorine of the source spreads into the system.

Source Quality

In junctions where water enters the network, this parameter is used to describe its chlorine concentration. For example, you can model a drop chlorine injector with a discharge of 0.001 l/s and a source quality of 100 ppm. In Chapter 6 chlorination will be described in detail.

Entering data for pipes

We are not going to talk here about the pros and cons of each type of pipe. When modeling in a development context you need to consider mainly two groups of pipes:

a) **Metal pipes.** They have greater friction, they absorb chlorine, their diameter is decreased by build-up and scaling on the inside of the pipes and they are more expensive. They include galvanized iron and cast-iron.

b) **Plastic pipes**, are smoother, don't absorb chlorine, and they accumulate less build-up and are cheaper. They include PVC and polyethylene (PEAD).

The introduction of data for pipes is much easier. The first three parameters are given by EPANET automatically. Although you can change the ID of a pipe, this is a big waste of time.

When you make modifications, the junctions' IDs keep changing. You can change the names when you finish with the Rename-IDs tool.

Property	Value
*Pipe ID	488.11
*Start Node	122
*End Node	123
Description	
Tag	
*Length	57
*Diameter	150
*Roughness	140
Loss Coeff.	1.2
Initial Status	Open
Bulk Coeff.	
Wall Coeff.	
Flow	#N/A
Velocity	#N/A
Unit Headloss	#N/A
Friction Factor	#N/A
Reaction Rate	#N/A
Quality	#N/A
Status	#N/A

The last 7 are read once the model has been analyzed. Until then, they will show up "#N/A".

Length

With a calibrated image, the length will be determined by EPANET in the auto-length mode. If you do not have a background image, you will have to enter the lengths pipe by pipe from the topographic survey.

Diameter

Metal pipes are specified by the internal diameter. A 25 mm pipe has 25 mm of inside diameter, and this is the diameter that is entered when modeling them. In contrast, plastic pipes (PVC and PEAD) are classified by their outside diameter.

The inside diameter is calculated by taking the outside diameter minus the thickness of the wall. **When modeling use only the internal diameter**. To make things more complicated, the specifications vary from one manufacturer to another however in practice, it is acceptable to use an approximated integer, as it causes little errors, simplifies the task enormously and avoids errors. Use this table of approximated relationship between commercial or nominal diameters (DN) and inside diameters (ID):

DN	25	32	40	50	63	75	90	110	125	140	160	180	200	250	315	400	450	500
ID PEAD	20	26	35	44	55	66	79	97	110	123	141	159	176	220	277	353	397	462
ID PVC	21	29	36	45	57	68	81	102	115	129	148	159	185	231	291	369	--	462

Roughness

The roughness coefficient depends on the formula that is used to calculate the system hydraulically. The two main options are as follows:

a) If you use the **Darcy-Weisbach equation**, which is largely used in Europe, this coefficient is f and it takes values with decimals.

b) If you use the **Hazen-Williams**, as used in America, this coefficient is called C and it takes values like 100, 120, etc. The bigger the coefficient, the smaller the friction of the pipe.

I would recommend you to always use Hazen-Williams, because it is more intuitive. The coefficient has no decimals, and moreover, this coefficient does not depend so much on the diameter or velocity. Its detractors state that it is an experimental equation (Darcy-Weisbach is theoretical) and that it only works for water at room temperature, but…isn't that just what we want to calculate?!

If my argument has convinced you, and you are not modeling the hydraulic behavior of fish soap at 90°C, you must make sure that you have the Hazen-Williams equation selected (represented by H-W in EPANET). In order to do that go to >*Project/Analysis options*, and select H-W. [10]

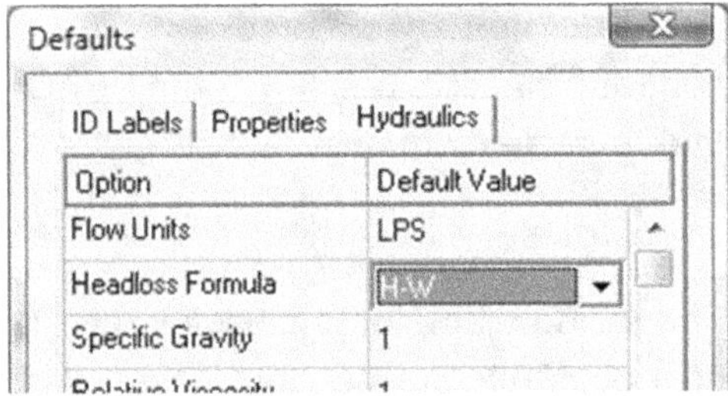

The value of C depends on material and state (and slightly the diameter), being bigger for plastic and new pipes. Basically for pipes of:

Plastic	140-150
Galvanized iron	120-133
Cast-iron ducts	100

Use the EPANET help for a more complete listing.

Make sure you don't use H-W C-coefficients with the D-W formula or vice versa. That would make your model throw negative pressures despite installing enormous pipe sizes.

Minor Loss Coefficient

Minor losses account for the energy losses produced by the turbulence in the flow as it goes through anything that is not straight pipe: elbows, valves, reducers, tees, etc.

---

[10] *Don't worry, in the United States they have been calculating networks like that for 100 years.*

In the image, the half-closed valve creates a conical turbulence on the outlet of the tap which is clearly shown on the wall reflection and against the blue cloth. Producing this whirl consumes part of the energy available in water. The turbulence created when changing the velocity and direction of the water flowing in the pipe as it is forced through a fitting produces the friction that uses up pressure.

In most cases, the losses are so small that they are not included in the models. Cases in which they should be included are:

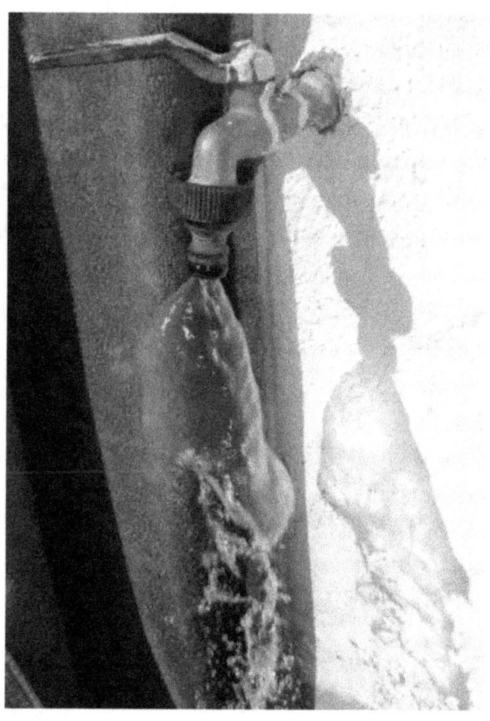

- Pumping stations.
- Internal building installations and diameters smaller than 1".
- Places where water flows at high speed.

One way of presenting minor losses is by adding an equivalent length to the pipe, that is to say, X extra meters produce the same friction as the accessories added. Even if this may seem quite practical at first, it presents two main disadvantages:

- Because EPANET would not be representing real pipe lengths, you could not take data directly from the program in order to determine the materials required. Mixing real and fictional lengths can generate a very confusing database.

- Water quality analysis is affected. When you add a pipe length, you affect travel times making them longer, however the reaction does not have in reality this extra time which produces false results.

To avoid these problems we use the loss coefficient. This coefficient (K) is characteristic of each accessory and it is related to energy loss in water head meters (H) with the formula:

$$H = K * v^2/2g \qquad v: \text{velocity and } g = 9.81 \text{ m/s}^2.$$

The coefficients given with EPANET´s help are faulty at the time of writing. Please use Table 2.6 of *Advanced Water Distribution Modeling and Management* or other source.

Be careful when you draw a pipe with an angle of 90°, for example, and want to take into account the losses as you need to include its value for loss coefficient. With EPANET, no matter how you draw the pipes, they are all straight and with no accessories. To illustrate this, you can see that the pressure is the same in both pipes with identical section and length, even though one of them is not straight. EPANET has not taken into account the great number of elbows and the losses needed in order to build pipe B.

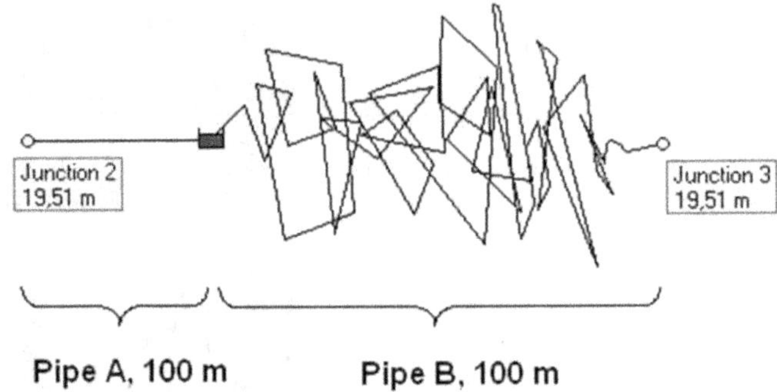

Initial State

Refer to valves further in this section. It basically describes if they are open, shut, or only allow flow in one direction. This is the way to incorporate shut-off and non-return valves.

For **Bulk and Wall Coefficients** see chapter 6.

Entering data for a reservoir

Here you have the most common parameters used in a development context. For others, please refer to the EPANET User's Manual.

Elevation or Total Head

This is the height of the water level. In some cases it can be tricky to determine.

Besides a river or lake, the most frequent case in development is a **borehole**. As you pump from an aquifer, the water level goes from rest (static level) to a much lower one (dynamic level) as is shown in the image. Broadly, a depression cone is formed, similar to the one that is created when pulling the plug out from the sink. The bigger the discharge, the greater the depression cone is. This cone makes the water level much lower inside the borehole and this is the level that needs to be entered. For example, if the borehole cap is at 30 m height, and static water level is 10 m under the ground, and when the pump is activated the equilibrium is reached at 50 m from the cap, the level of this "reservoir" will be 30 m – 50 m, and so -20 meters.

Reservoir 1	
Property	**Value**
*Reservoir ID	1
X-Coordinate	-1341.10
Y-Coordinate	9448.42
Description	
Tag	
*Total Head	60.40
Head Pattern	
Initial Quality	
Source Quality	
Net Inflow	#N/A
Elevation	#N/A
Pressure	#N/A
Quality	#N/A

Another common error occurs when entering pumping head of the pump. If the dynamic level is 20 m above the pump, the pump does not do any work to move a fluid inside another fluid of the same density, that is to say, it does not matter what depth the pump is installed below the dynamic water level. Only when raising water over air is an effort made. To visualize this concept, put a bag of water into water and see that it only requires effort to move it when trying to take it out of the water.

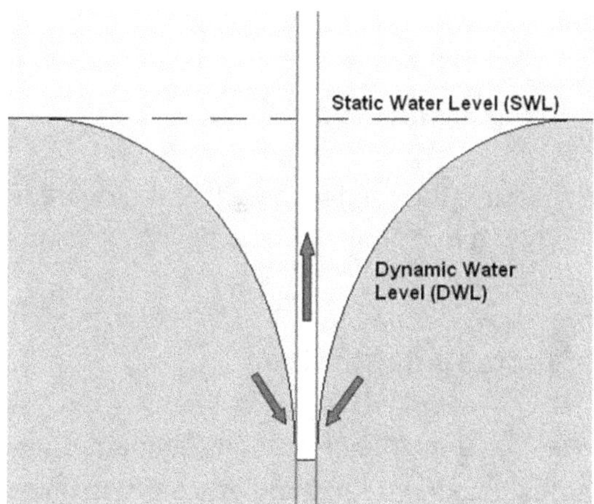

Initial Quality and Source Quality

See chapter 6.

Entering data for a tank

We are going to enter one of the most common tanks in emergencies, an Oxfam T95. It is a corrugated tank, which is 3 m high and has a diameter of 6.4 m. The outlet is 20 cm above the floor and the overflow outlet is at 2.8 m.

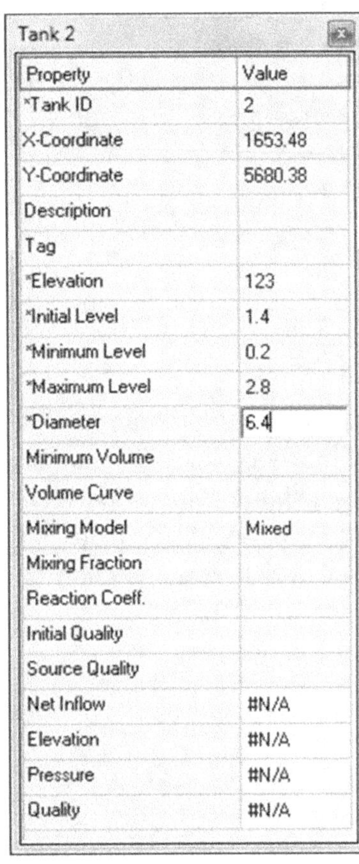

Elevation

This is the height of the tank base, for example, 123 m above sea level. It is used as a reference for the data that follows.

Initial level

This is the initial water level in the tank. If it is half full: 1.4 m.

Minimum level

This is the relative height of the water outlet for distribution. Do not mistake it for the drainage level or the fire reserve outlet. On the T95, it is 0.2 m.

Maximum level

This is the relative height of the overflow. In our case this is 2.8 m.

Diameter and "rounding" of rectangular tanks

EPANET assumes that all deposits have a circular section. Although tanks are modeled as if they were cylindrical, in reality most tanks are rectangular…argh! In order to overcome this difficulty, convert to the "equivalent diameter", which is as simple as finding the diameter of the circle that has the same area as your tank.

It's best to explain it with an example:

An 8 m x 12 m tank has an area of 96 m². What we need to do is to find the diameter of a 96 m² circle. The equation that we get is:

$$96m^2 = \frac{\Pi * D^2}{4}$$ The diameter to be introduced is then 11.05 m.

The generic formula is, with A and B being the width and the length: $$D = 2\sqrt{\frac{AB}{\Pi}}$$

Minimum volume

Ignore this. This is to help calculate mixing with the residual volume when the tank has uncommon shapes.

Volume curve

This is used for tanks with a variable diameter, in order to relate volume with height. Tanks with variable diameters are conical, spherical, etc.

Mixing model

There are four mixing models of which the first one is the most common:

1. **Complete.** Mixing is complete and instantaneous. It is ideal for systems that fill a tank completely and empty them afterwards. This frequently happens in a developing context because the easiest way of chlorinating the water is to fill the tank completely, dose the chlorine and wait for it to act before it is opened for consumption.

2. **Two compartments.** This is uncommon. Refer to the EPANET User's Manual for further details.

3. **LIFO.** This stands for Last in First Out and assumes that water is not mixing. The last water to enter the tanks is the first to leave. It can be used to model treatment tanks.

4. **FIFO.** This stands for First in First Out. Refer to the EPANET User's Manual.

Mixing fraction

This is only needed for the 2 compartment mixing model. Refer to the EPANET User's Manual.

For **Reaction coefficient, Initial quality** and **Source quality,** see chapter 6.

Valves

As we stated earlier, shut-off and non-return valves are modeled as a property of the pipe. They are not represented as an object, but as a property within the pipe. In order to close a pipe or install a check valve you need to change the property *Initial state*: open, close or check valve, which only allows flow in one direction.

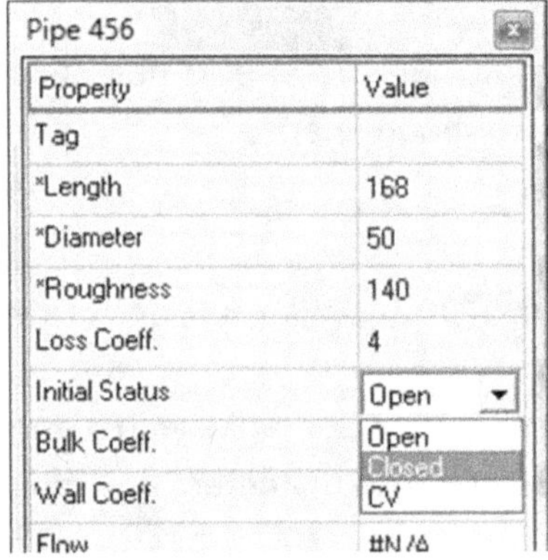

Other valves, and this is my personal opinion, should be avoided in most cases in a development context. The reason is that these valves are delicate and expensive. A 200 mm pressure sustaining valve costs over 3,500 €. This high cost and lack of local experience in using these valves plus lack of availability in local markets makes replacement very unlikely. I know that you will agree that water pumps are easily available and widely known…however, I have seen many abandoned water systems because the local community could not afford the cost of a new pump!?!

So avoid the following valves:

- PSV (pressure sustaining valves). Prevent pressure upstream from going under a certain value.
- PRV (pressure reducing valve). Reduce pressure downstream to below a certain value.
- FCV (flow control valve).
- Throttle valves. Valves that are partially open, i.e. butterfly valves.
- PBV (pressure breaker valve). Force a pressure drop.

If after all you decide that you need one of these valves, simulating them is easy and you can handle it by using the EPANET User's Manual.

Modeling a pumping station

Don't add pumps at this stage even if they are included in the network. This is because the pipes are dimensioned first and the network stabilized. Once this is done the suitable pumps are added so they work for that specific network. Wherever the pump goes, put instead an imaginary reservoir at a certain height and proceed as if it was a gravity system. Work with the pipes until you can lower the tank as far down the hill as possible. The height of the hill will be the water pumping head (the height the pump should be able to lift), at it will be more economical to pump to lower altitudes. Look at the image of the section Skeletonization further ahead, and this is exactly what has been done, and why there are so many reservoirs.

Here is an example highlighting some points:

- The type of pump (centrifuge, horizontal, submersible) does not affect the simulation if the intake of the pump has enough water pressure so that the pump does not suck in air or produce it (google NPSH for more info).

- If there are big demand variations, several pumps of different sizes are used instead of a bigger one so that they can be turned on and off depending on the required discharge of the moment.

Modeling a submersible pump for a borehole

Let's design a very simple network in a flat area, let's say in Eritrea. The water is taken from a borehole and is directly distributed to four points.

1. Start drawing this network without connecting the borehole (it will be modeled as a reservoir).

2. Enter the following data: 0 m for elevations, 100 m for pipe lengths and, to start, 200 mm for pipe diameter.

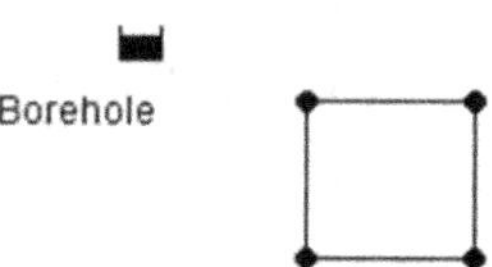

3. Assign the maximum pressure you want for the system to the reservoir, above the height of the lowest point. Let's say that I want 3 bars, equivalent to 30 m of head. Because the lowest height is 0, the reservoir elevation is 0 m + 30 m = 30 m.

4. Join the reservoir with a pipe.

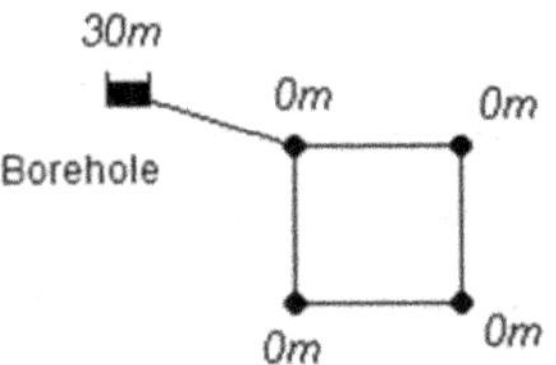

5. Now we need to follow a process that will not be explained here but will be covered in future chapters. Basically you would start changing pipe lengths and diameters until getting the optimal solution for the network, which will have allowed us to lower the reservoir to 25 meters. This is the result of the process:

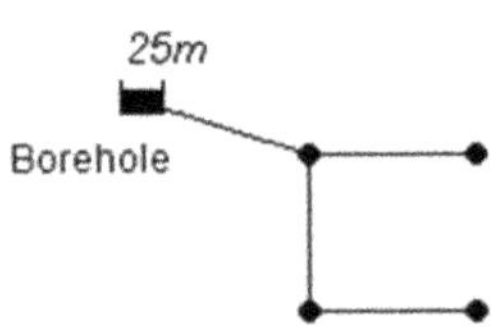

If the borehole is not drilled or the pumping test has not been done, you cannot know how deep the water will go when pumping (**dynamic level**), or what discharge the well will provide. Because you do not have enough data yet, you need to leave the model like this for now.

6. Now we make changes to accommodate the pump. Remember that it is a borehole; when we start pumping the water level goes down. Place the reservoir that you are simulating at a height that equals the height of the borehole top minus the dynamic level. This means that if you had the cap of a borehole at an elevation of 5 m, and when you pump the water, the water level goes 35 m below that point, the height you need to introduce for the reservoir is 5 m – 35 m = -30 m.

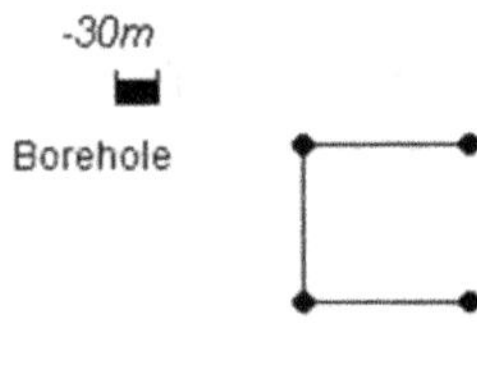

Watch out, remember that 35 m is not the depth at which you need to install the pump, but the depth the water table reaches in the drilling test. Now delete the pipe that connects the borehole to the junction on the network.

7. Select a pump that provides the suitable discharge as you will see in the following chapters. The pumping head or pumping height is the sum of:

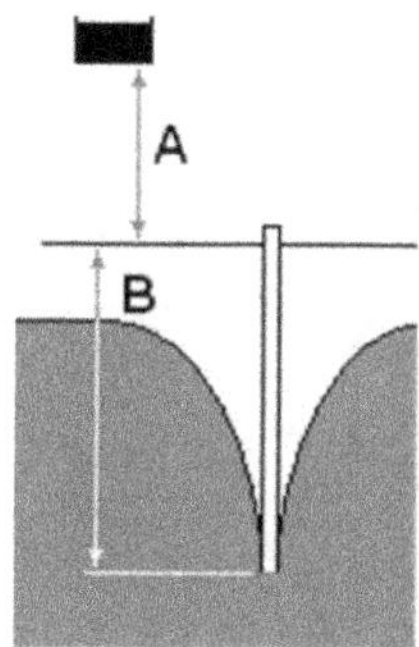

25 m   (A, calculated in point 4)

30 m   (B, from the dynamic water level to the top borehole, as calculated in point 6)

The pumping height would be 55 meters, which is the total height from the dynamic level up to the hypothetical reservoir that we had placed in point 4. This hypothetical elevation simulates the resistance of the network to receive the water plus the residual pressure.

8. Draw the pump remembering that the pumping direction is the same as shooting from an imaginary cannon. You will learn how to add the pump properties in the following section.

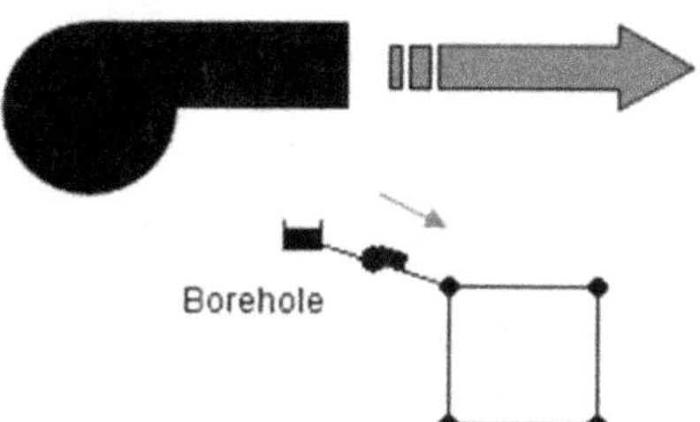

9. There is no need to put a pipe as the pump assumes it automatically, however, it does not include the parameters of that pipe. If you think that can affect the systems behavior, for example because there is a great distance, add an intermediate junction and assign the properties of that pumping main.

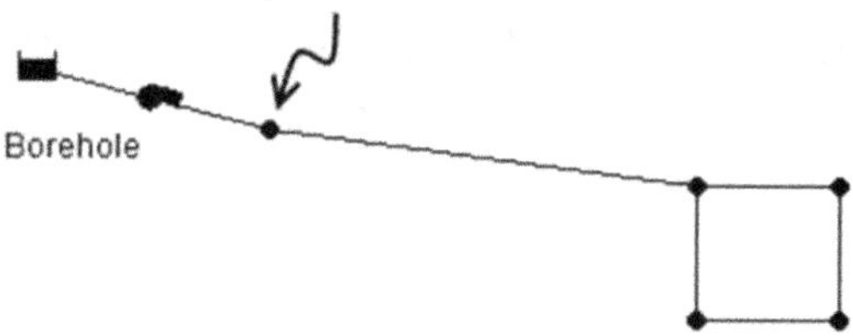

Adding a pump

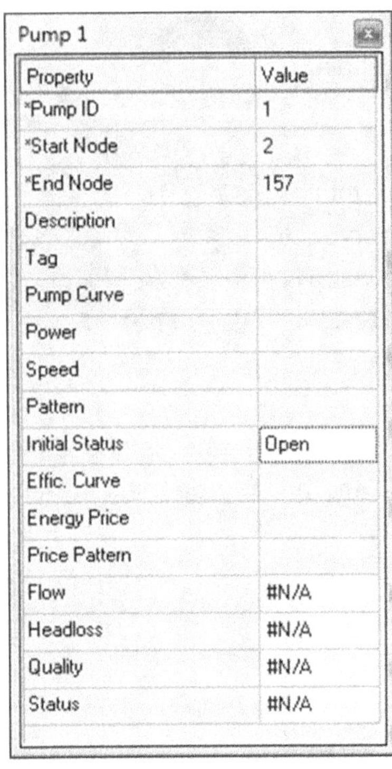

Start Node

This is the junction from which the pump takes water.

End node

This is where the water is taken. Remember:

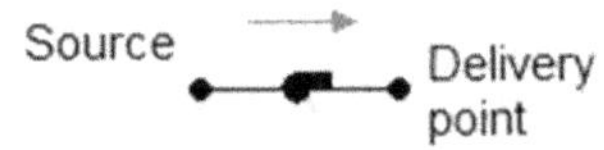

Pump curve

This is a curve that represents the pumping head pump.

If you are not familiar with this, or you're not entirely sure what it is, this is a good chance for you to look up a pump book or google for it. On the right you can see the three curves that define a pump's behavior: the pump curve, the performance curve and the suction curve for non-submersible pumps.

In this parameter you enter the name of the curve that you have constructed using the manufacturer's manual. In order to make the pump curve follow the instructions that were given to make the tank volume curve making sure that you choose *Pump* as the *Curve type* and that the pairs of numbers are the height with respect to the discharge.

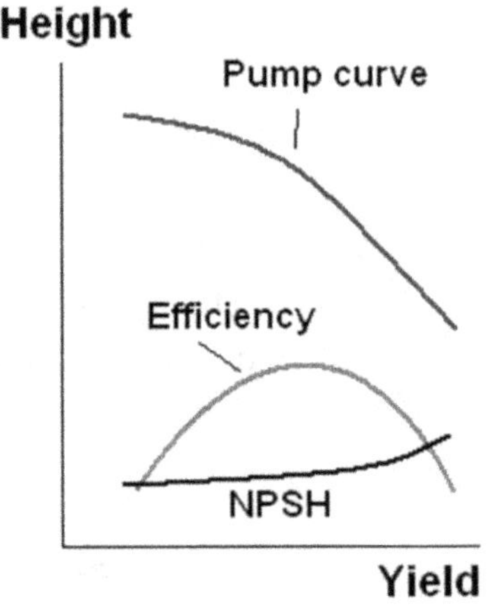

Power (kW)

This is used when the pump curve of the pump is unknown, and for other cases is left blank. In the design phase this is not going to be needed because you are going to choose the pump yourself, but it is very useful for boreholes that already exist, when you do not know which pump is inside.

Speed

EPANET can also model variable speed pumps, like solar pumps, using this parameter.

For **Efficiency curve, Energy price** and **Price pattern,** see chapter 8.

Modeling a pump

The actual type of pump (centrifugal, horizontal, submerged) modeled in EPANET does not matter much, provided that the pump inlet receives enough water at a high enough pressure to avoid cavitation.

Modeling a particular pump

This is the case where you know exactly which pump is already or will be installed. As an example, let's model the Grundfos CH 8-30 pump pumping water from a shallow well to a 4-node network 2 kilometers away.

1. Draw a reservoir and place a dummy node nearby.

Pay attention because this is potentially disastrous. For some reason, EPANET draws the pumps with a pseudo-pipe included that doesn't allow you to enter parameters. It will not take into account, for example, the head losses inside the pipe. You could build a network and not get water because it didn't consider these losses. To avoid this, **always add a dummy node from which to draw the actual pipe, enter its parameters and take into account the head loss it produces.**

2. Draw the pump the way you draw pipes, by clicking on the reservoir from which it draws the water and then on the dummy node. In EPANET, the pumps have the shape of a cannon. Make sure you **draw the pump so that the cannon points in the same direction that the water will flow**:

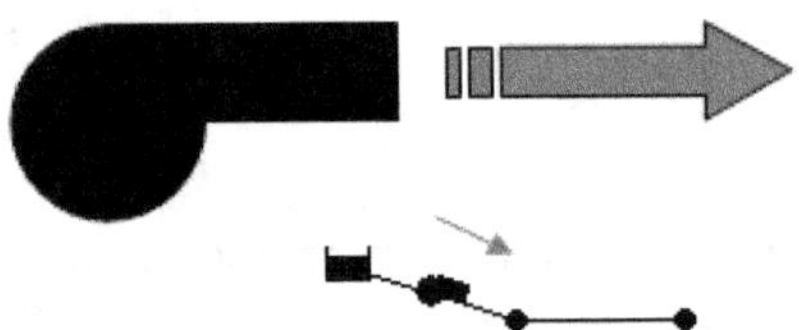

3. Add a 2 km pipe connecting the dummy node to the network:

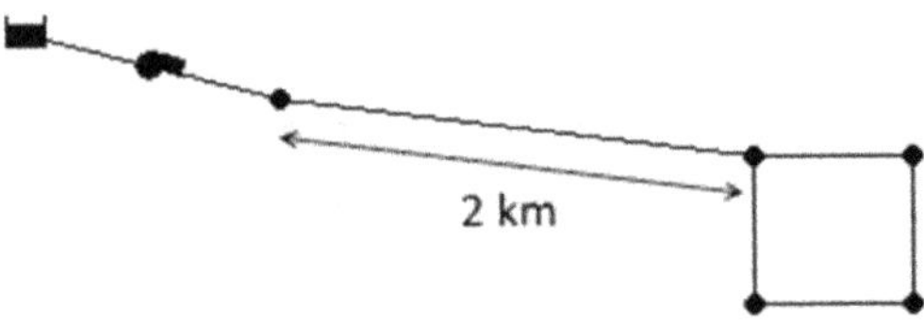

4. Next we enter the pumping curves in >*Browser/Data/Curves*. A dialog box appears. Enter a name in the *Curve ID* field, for example CH 8-30, and select *PUMP* in *Curve Type*. You can add more details about the pump in the *Description* field:

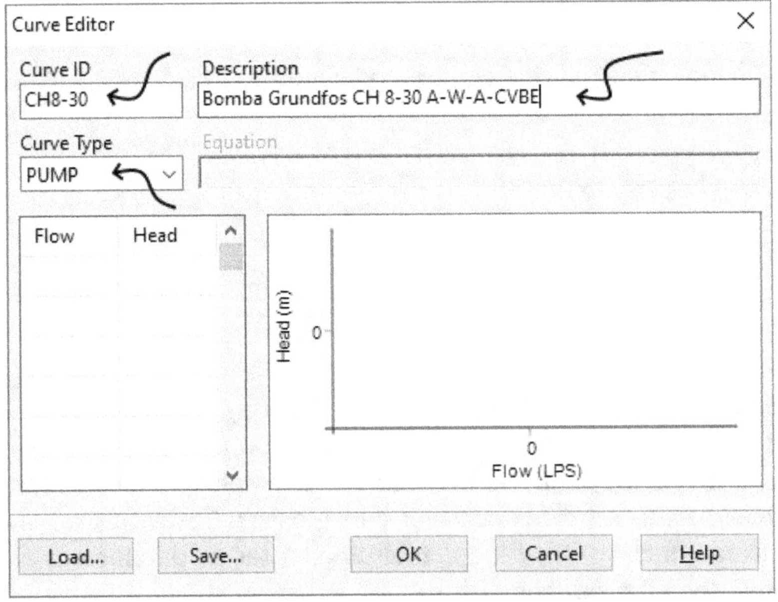

5. Using the curves provided by the manufacturer and identify 4 points at equal distance from each other on the segment of the curve where the pump efficiency is higher.

In this example we will take the following points:

(4 m³/h , 28 m) ; (6 m³/h , 25 m) ; (8 m³/h , 21 m) ; (10 m³/h , 16 m)

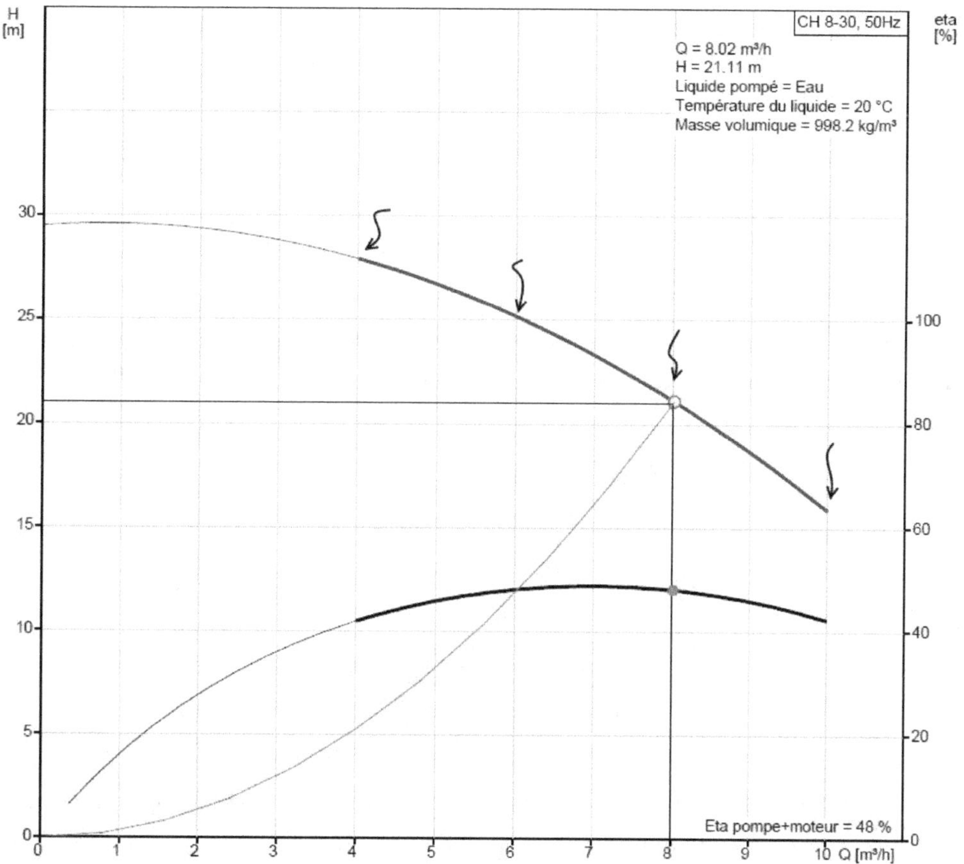

Once converted in l/s, the units EPANET is using, we obtain:

(1.11 l/s , 28 m); (1.66 l/s , 25 m); (2.22 l/s , 21 m); (2.78 l/s , 16 m)

Enter these values **in ascending order of flow** in the *Flow* and *Head* columns of the dialog box:

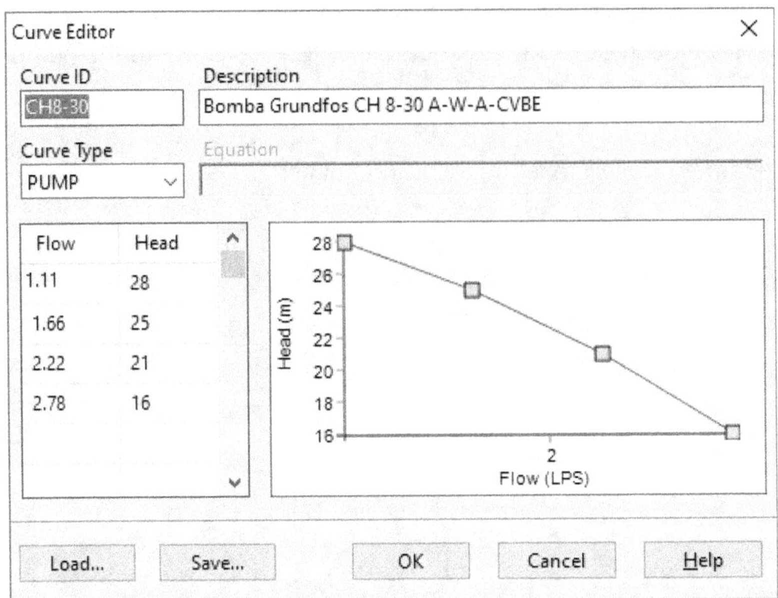

If you have enough information, it is very important to **build this curve from at least 4 different points**. Notice how the curve created from a single point is different from the one above:

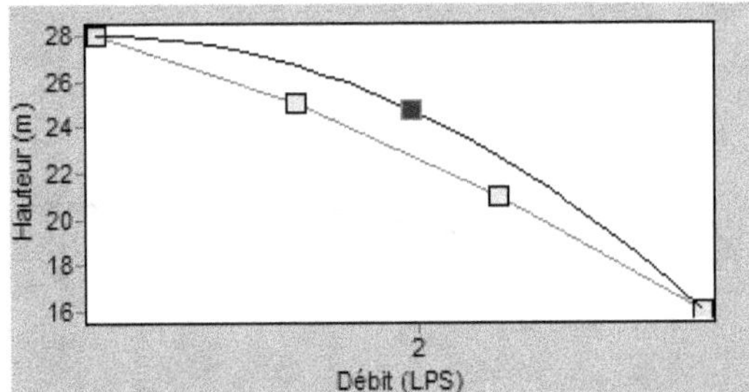

6. Finally, assign the created curve to the pump, click on the pump and enter the curve name in the *Pump curve* field, in this example CH 8-30.

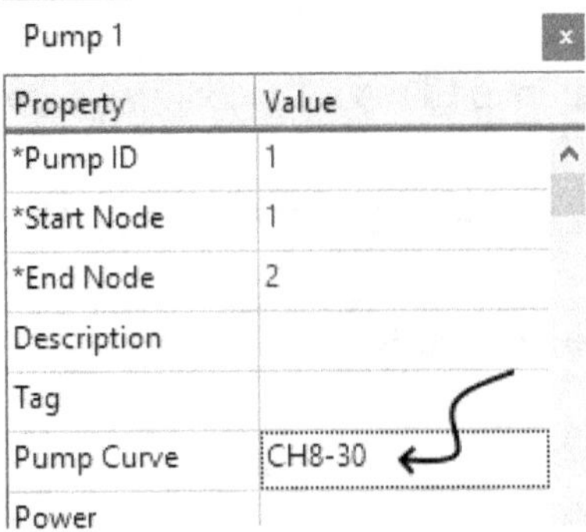

If you do not know which pump will be installed

1. Create a pump curve from a single point by entering the desired flow rate and the height difference between the source's and the destination's water surfaces
2. To size the pump, modify the flow rate and head until a satisfactory solution is found. Or simply calculate the operating point by hand.
3. Once these values are established, find the most efficient pump at the working point found from the manufacturer's literature.
4. Create a curve for this pump using at least 4 points taken from the manufacturer's literature as explained above and check that the system is working properly.

Modeling a borehole

To model a borehole, proceed as in the previous section, except that the height of the reservoir is calculated differently. When water is pumped from an aquifer, a depression cone is created. The height of the cone will be equal to the height of the well head minus the height of the dynamic water level. For example, if the well is located at an altitude of 100 meters and the water level after pumping drops to a depth of 70 meters from the borehole head, then the water level will be 100 m - 70 m = 30 m. This is the value you enter in *Total head*.

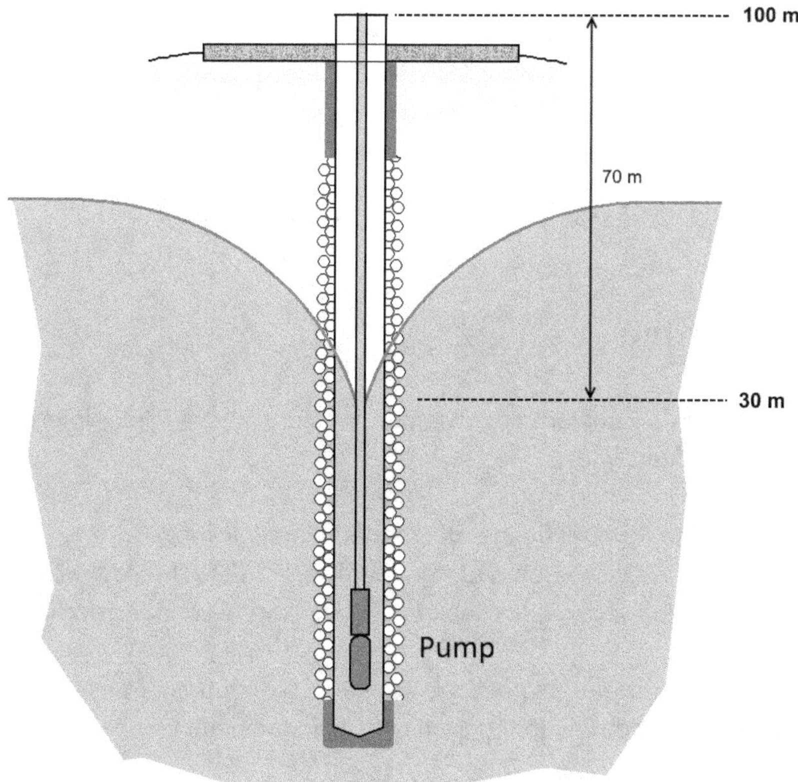

A common error is to enter the pump's elevation. It doesn't matter how deep the pump is installed. Raising water in water is effortless because they weigh the same. The pump only starts working when it pumps out of the water. To visualize this concept, imagine that you submerge a bag of water in a swimming pool. Only when you try to pull the bag out of the water will you feel its weight.

Don't forget to **consider the rising main inside the well**, that is, the vertical pipe that goes from the pump to the head of the well. Since the diameter of the pipes are usually different inside and outside the borehole, you probably have two different lengths of pipe.

This diagram shows a network where the pump, installed at a depth of 175 meters, is connected to a first tube of 75 mm in diameter which increases to 150 mm and 2200 m long when it comes out of the borehole:

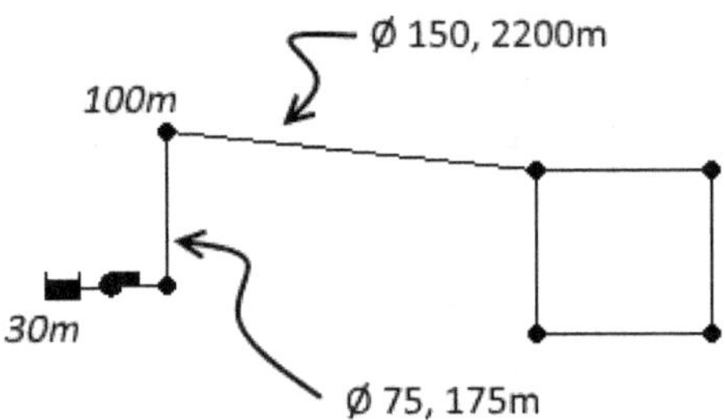

https://youtu.be/tv-RBMr2VHc   (Modeling a borehole)

Modeling a spring

The key is to achieve a constant flow without pressure at the inlet. Unfortunately, if you connect the network directly to:

- A reservoir, the flow will be that demanded by the network.
- A node with a negative demand (for example, - 2 l/s). The flow will then be exactly 2 l/s, but will forcefully inject those 2 l/s into the network, distorting the pressure results.
- A flow control valve is impractical, does not give you information about the water that overflows, and fills the program with dialogues that you have to click away.

The solution is to install a node with negative demand (A), a reservoir that acts as a sink (C), and a connection node (B), as shown in the diagram on the right. The spring is represented by the set of elements included in the cloud. To model it, the procedure is as follows:

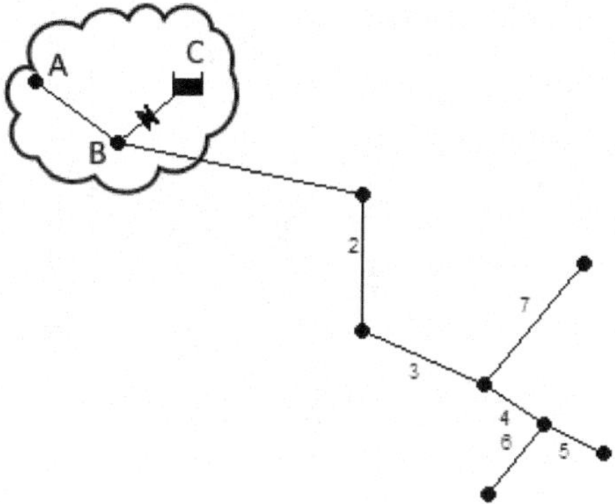

1. Draw the nodes A and B, and the reservoir C. These three elements must have the same altitude, that of the spring you are representing, for example, 326 m. Enter the flow rate of the spring as a negative quantity in the *Base Demand* of node A. If the source has a flow rate of 2.34 l/s, enter "- 2.34 l/s".

2. Connect nodes A and B with a dummy pipe (one with a very large diameter and a length that does not affect the behavior of the network, for example, 1 meter in diameter and 1 meter in length).

3. Draw another dummy pipe to connect nodes B and C. In that order! And add a non-return valve in the *Initial status* field. The water must be able to enter but not leave the reservoir, preventing the reservoir from discharging into the network.

[▶] https://youtu.be/-KU43lo3ZAs   (How to model a spring)

Modeling a pressure rupture tank

In mountainous regions, it may be necessary to dissipate the pressure that builds up due to the difference in elevation. In poor contexts, pressure reducing valves that would do this job are often replaced by break pressure tanks (BPT), which are much more robust. The following diagram shows a BPT. The principle is pretty simple, provide a big opening releasing the pressure into the air without generating a leak, so that the pressure in the network falls to 0.

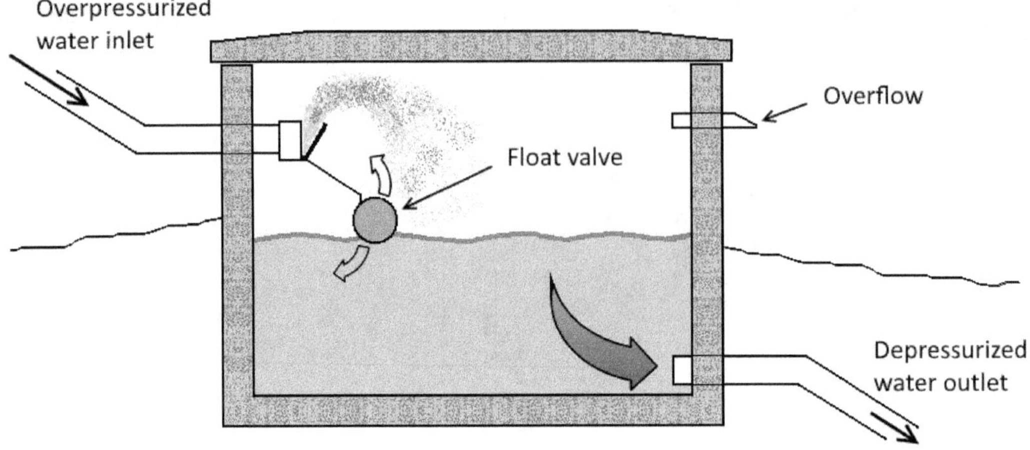

To model a pressure rupture tank:

1. Find out the location and elevation that places the pressure in an optimal range in the network. Then draw two dummy nodes and give them that elevation. They represent the entrance and exit of the tank.

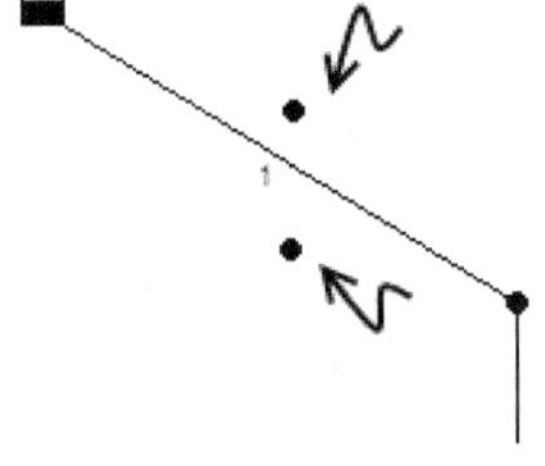

2. Draw a pressure reducing valve (PRV) and enter a value of 0 in the *Setting* field. The diameter should be at least equal to the diameter of the pipe. Make sure you draw it in the expected direction of flow to avoid random errors from EPANET.

3. Divide the existing pipe into two sections, as shown in the diagram, specifying the length of each pipe according to the location of the BPT.

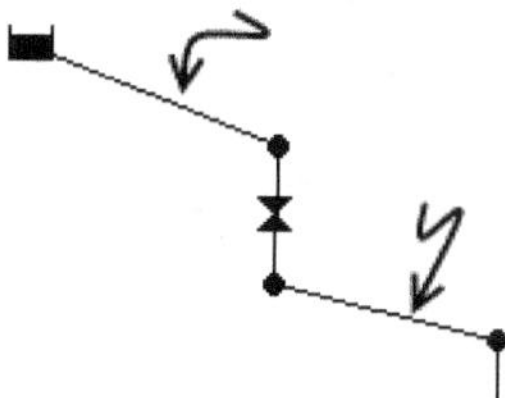

In the example below from a system in Lesotho, you can see that water enters the tank with a pressure of 83.69 m and exits with a pressure of zero.

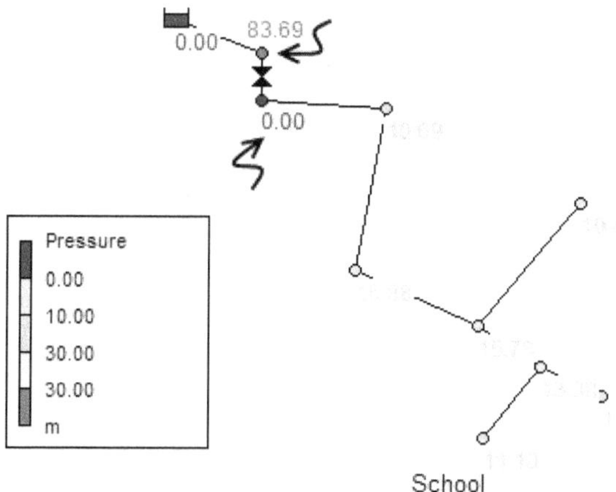

https://youtu.be/NwpRnu09X-Q   (How to model a break pressure tank)

Skeletonization

If a model considered absolutely every component it would be very laborious and expensive to build, impossible to maintain and difficult to interpret. Skeletonization is a process which eliminates the pipes that have little effect on the network's behavior. Both images below are the same network, one with all the data, and the other one simplified to analyze the loop behavior. In this extreme simplification it has been considered that all the deleted elements do not contribute much to the loop's analysis.

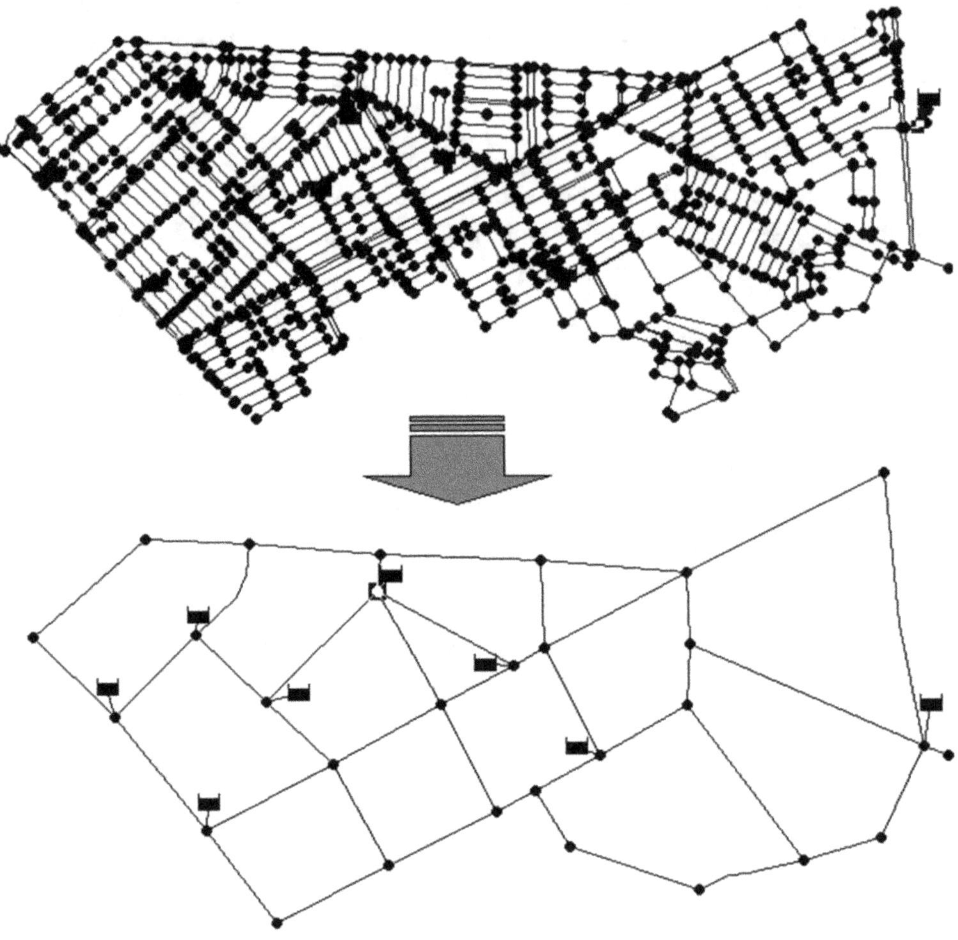

In developing contexts, models are relatively simple and only the really unnecessary details are eliminated. For example, these three things don't need to be modeled in detail:

- House connections. As you will see in the section on Demand allocation, the total demand of all houses in a reach of pipe is concentrated in one or two nodes on each end.
- Interior installations of each user are replaced by a node.
- Detailed connections of fittings are ignored, as in the image below.

Only with these principles, the network is simplified considerably and I believe all the necessities of common systems are covered without risks. If you need to simplify your network any further, check out some books, adapt the model to the changes and see what happens. Alternatively you can use specific expensive software like Skelebrator, or a free beta program, like optiSkeleton.

In many cases, skeletonization does not mean information loss as existing parts will include that information. For example, the group of accessories in the image below will be simply transformed into three pipes with minor losses included.

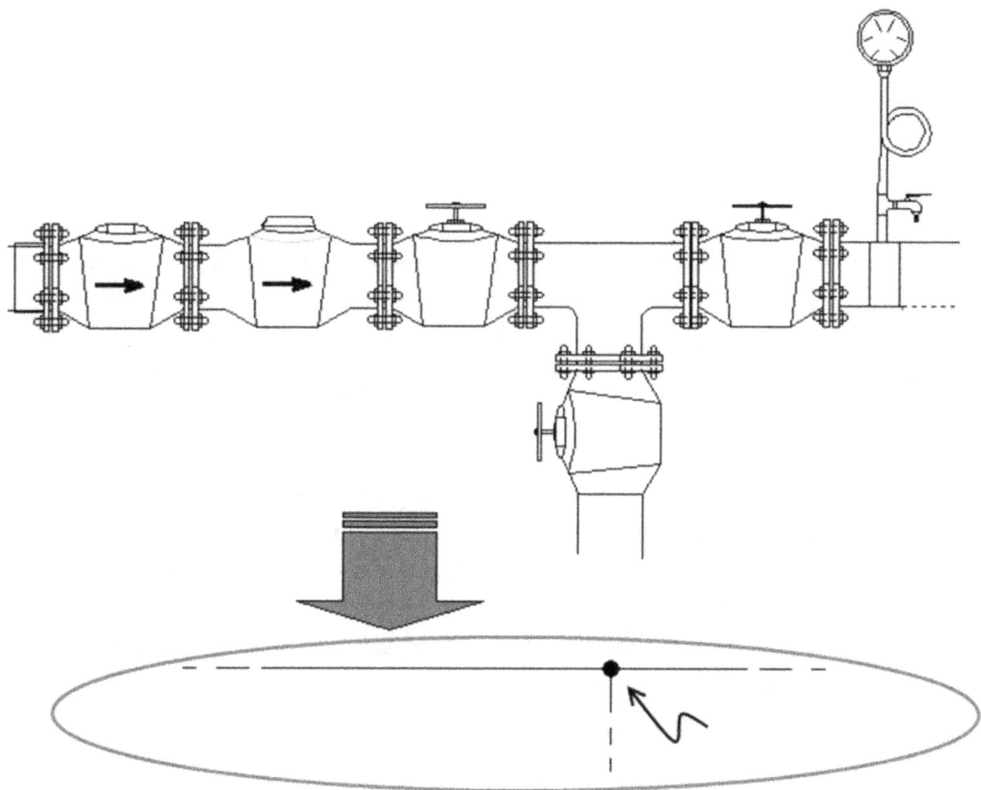

Avoiding messing up spectacularly

I have left this for the end of the chapter in the hope of making it more striking. The temptations to succumb to these two points are tremendous, because they result a priori in much cheaper systems, but without any future viability. If you don't want to pull your hair out, respect these recommendations:

Minimum diameter: 63 mm

With the exception of house connections, **do not install pipes smaller than 63 mm under any circumstances**. You will see that this is also important to facilitate the calculation of systems later and almost fundamental to use calculation programs, but the reasons here are different.

Small diameter pipes are very treacherous. They block very easily with air or sediments, are very sensitive to small variations in diameter, do not allow future extensions or corrections, do not tolerate the slightest error or unexpected circumstance and do not offer fire protection. If, as in a certain project, 5 km of 25 mm pipe are installed, you know that after a few months they will be blocked, and then try to find out in the middle of the rainforest where exactly it has got stuck!

Minimum pressure rating: PN10

Do not install pipes rated less than 10 bar, *regardless of the pressure that comes out in the calculation*. And 10 bar is already a lowball, most water utilities install PN16 directly, because they know that the costs of repairing breaks and damage to neighboring infrastructure are much higher. Install PN16 or directly metallic pipes under roads and other critical places.

The reason is that pipes do not only withstand internal water pressure, they also withstand ground pressure, thermal change stresses, the weight of vehicles (also that of giraffes, water buffaloes and other heavy animals), and when installed and transported, they end up having tool marks that weaken them. Another fundamental reason is that the pipes with less rating, for example PN6, are for agriculture, and although they are suitable for human consumption as they will tell you, they are designed for much shorter *laboratory* lifespans, 25 years, than those for human consumption supply, 100 years.

But 25 years is an illusion, if you install those pipes, you will have problems after a few weeks or months.

CHAPTER 5

Loading the model

Introduction

Loading a model is applying the "load" of the consumers on the network. It is a fundamental step but comes with less solid information. Avoid taking an overly cautious approach, pay attention and calmly accept that not everything is going to be based on solid data. This is the territory of assumptions and it is important to work with the correct ones.

The model is loaded to work adequately at the hour of the day, of the day of the week, of the month of the year when the future population will consume the most, that it to say, it's **designed for the peak consumption point.** This way, the networks are slightly over-designed, but the advantages definitely deserve a slightly larger investment. The philosophy behind this is that if the system works in the worst case scenario, it will work in the others. Emergency and small systems are calculated with other approaches explained later.

Demand is consumption of an individual, business, factory, etc. To make data comparable and easy to modify, the demand is divided up half-hourly. Multipliers and coefficients will be applied to take into account the hourly, weekly, monthly variations, etc.

Besides the value itself, **the spatial distribution of the demand is vital.** At the end of the chapter, techniques to assign demands to each junction are described.

> https://youtu.be/HaUFU9Pm1Uc   (Design overview)

Two aspects with strong social implications are fire demand and long design periods. Both require big financial investments which could perhaps be better used in other community issues, such as vaccination or lighting. It's about finding a compromise that avoids blindly following Western design standards.

Demand is a self-fulfilling prophecy. If you have designed a network correctly, you have planned it for the community that is going to receive it. Make sure that the quantity per person is adjusted to the population. If you don't plan your network properly, it is the population that will have to adapt to the network and not the other way around. The network could ultimately disrupt people's daily lives, school schedules and so on.

Current population vs. Future population… ¡ouch!

When a network is planned, it takes into account the population after a certain number of years, usually 30. To estimate the future population, formulas are used to project it over time according to its growth rate to give a graph as shown in the image. If in 2001 the population is 90,000, in 2036 according to the geometric formula it will be 230,000:

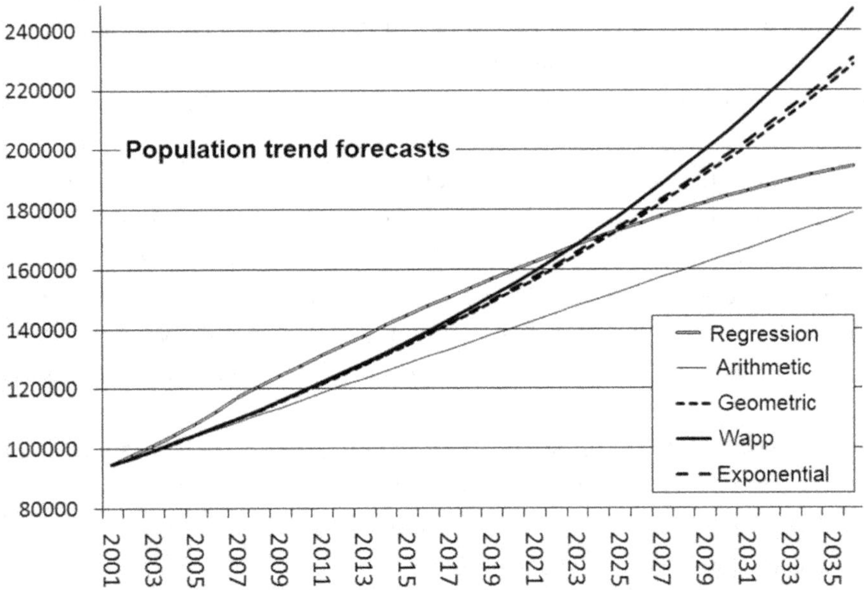

But things are not so simple, and here we begin to have difficult questions, for example:

- Is it right to pay attention to the need of the population in 30 years' time if it means neglecting some of the needs of the current population?

- If the network is built without taking into account the future population it will be obsolete before it is paid back, but if it is built too big… will it not run the risk of the community not being able to maintain it?

There is not a clear answer to these questions, but there are ways to deal with these matters. Some ideas are below:

1. **Design easily enlargeable networks.** This generally involves making big loops rather than branch layouts and not installing pipes less than 100 mm. As you see in the graph, the installation expenses are proportionately very big for small diameters therefore inevitably it ends up being preferable to install bigger pipes at very little additional cost.

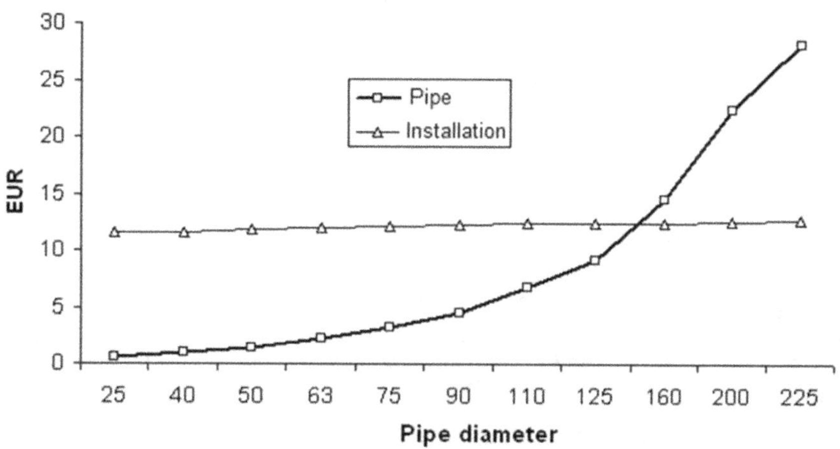

HDPE pipe cost vs., excavation, sand bed and installation cost, Afghanistan 2002

2. **Use plastic materials** in networks with a high possibility of growth. The price per increase in diameter is negligible compared to metallic ones. Once you have assumed the cost of excavating trenches and filling material, to install bigger pipes is not too expensive (see chapter 8).

3. For urban settlements you can use **density limits** in place of projections. If the population in the surrounding areas cannot tolerate more than a certain number of people per km², the population served will be the result of multiplying the working area by the density, for example:

$$8 \text{ km}^2 \times 1650 \text{ people} / \text{km}^2 = 13{,}200 \text{ people.}$$

4. **Plan the construction in two phases.** Using billing, you can determine how many years the system needs to save the necessary money to be able to finance a later enlargement (second phase). A very simplified example: if the net income of a water company is 10,000 €/year and it takes an investment of an additional 120,000€ to adapt the network for the needs of 30 year growth, design for at least twelve years, but better with a cushion of a few years.

Projection formulas

Arithmetic: $P_f = P_o \left(1 + \dfrac{i * t}{100}\right)$

Geometric: $P_f = P_o \left(1 + \dfrac{i}{100}\right)^t$

Exponential: $P_f = P_o * e^{\left(\dfrac{i * t}{100}\right)}$

P_f , Future population

P_0 , Current population

i , growth rate in %

t , time in years

e , Euler's number ($\approx$2,718)

This table, taken from the Bolivian Standard NB689, contains recommendations for the application of each method:

Method	Population (people)			
	Up to 5,000	5,000-20,000	20,000-100,000	> 100,000
Arithmetic	X	X		
Geometric	X	X	X	X
Exponential	X [2]	X [2]	X [1]	X
Log curve				X

[1] Optional, recommended

[2] Subject to justification

https://youtu.be/HaUFU9Pm1Uc   (Design overview)

Determination of the total daily demand

Next, and during a few sections we will see how to construct the demand that EPANET will utilize. You will only be able to enter values for demand when you arrive at the end of the section Global coefficient. Once you have arrived there you will be able to enter the parameters:

1. Base demand.
2. Modulation curve.
3. Demand type.

There is no quick recipe for determining the demand of any population. It varies a lot from one population to another and from certain conditions to others and you will need, above all, common sense to distribute it spatially. If there is another system close by you can get an idea, but watch out, the existing data doesn't tend to be correct. In emergencies, for example, the data is open to manipulation in order to sweet-talk the donors. In normal conditions it is extremely unknown and generally based on how much is produced, not how much is consumed. By way of orientation these are the **minimum standards** with which to work. I stress that they represent what arrives at the people, not what is produced in the boreholes, well, etc.:

Minimum consumptions (l/u*day)	
Urban inhabitant	50-100
School	5
Outpatient	5
Hospitalized patient	60
Ablution	2
Camel (once a week)	250
Goat and sheep	5
Cow	20
Horses, mules and donkeys	20

The cost of this quantity must be bearable by the families (maximum 3-5% of family income).

One area of consumption to investigate and avoid nasty surprises is small kitchen gardens. A typical garden consumes nearly 5 mm/m^2. Remember that 1 mm/m^2 is the same as 1 l/m^2, and a small kitchen garden of only 20 m^2 consumes 100 liters per day. If the custom of having kitchen gardens is widespread it can represent a considerable proportion of the total consumption.

In practice, the highest quantity of water will be set so that:

- It doesn't produce environmental problems (flooding, overexploitation, etc.).
- The people are willing to pay and at a cost that is adapted to the local economy.

Note that this is not the value that should be put as the base demand, it is only the starting point for calculating it.

So summarize, if I have 2 goats, 3 rural people and a donkey, the total daily demand is:

```
2 goats  x  5 l/goat            =    10 l
3 people  x 60 l/ person        =   180 l
1 donkey  x  20 l/donkey        =    20 l
                                -----------------------------
                                     210 liters per day
```

The effect of distance

The following has a dramatic effect on how the water system is conceived. Most water network design manuals assume that water is delivered directly to the homes, but this is not the case in many low or middle-income contexts. **You can plan for 100 liters per person, but if the water is far away, it will not be collected**. And by far, I do not mean 7 kilometers, I mean 30 meters!

If you plotted the amount of water collected in a house against the time needed for each trip to the water point, you would get a graph similar to the one below. In the first part (from 0 to 3 minutes approximately), water consumption falls very quickly from the taps in the house to about 20 meters away. Then comes a plateau where a constant amount of water is collected regardless of distance. From using the water almost on the fly as they need it, they go on to decide to make a certain number of trips. Maybe one bucket per person, maybe just enough to fill storage they have at home. In the third part (after 30 minutes) there is another drop. When the distance increases beyond a threshold, people simply collect the water that their time and strength allow them, which is usually a very low amount.

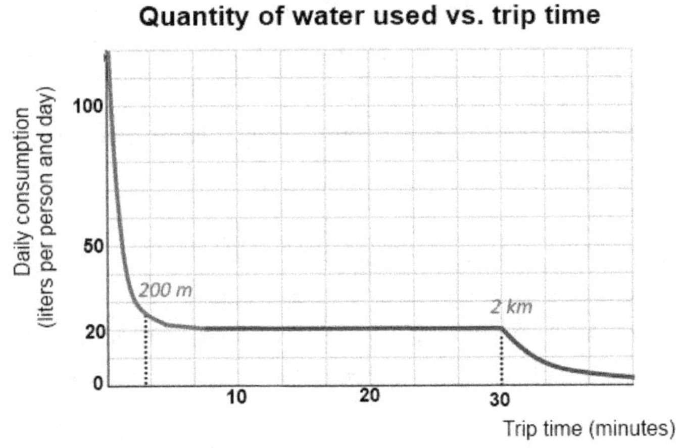

Quantity of water used vs. trip time

The last phase is a total failure, with no return on investment. The middle part is difficult to categorize. To make it visible think of a concept of ***getting ahead loss*** analogous to head loss. Even though they are collecting the same amount of water, **the longer the trip, the more social benefits of the project are lost**. Even if the same amount of water is collected, longer walks mean less time to attend to children, go to school or do any other high-value activity. Longer distances and bigger elevation differences lead to greater caloric expenditure. The link with food security often goes unnoticed, but in some settings **children spend as much as 25% of their caloric intake on collecting water**.

Multipliers

Consumption is not constant. People get up, go to bed, and are going to work at their individual whims and conveniences. To be able to represent the daily, weekly and monthly variations in an easy way we turn to the concept of multipliers.

A multiplier is a number that multiplies the average consumption of a population to give a real consumption in the timeframe considered (i.e. an hour). This way, if the average consumption is 100 liters per hour and the consumption is 50 between 12:00 and 13:00, the multiplier is 0.5 so that

100 liters	x	0.5	=	50 liters
Average demand	**x**	**multiplier**	**=**	**Real demand**

The use of multipliers allows calculations to be simplified and automated. Imagine that you want to see what happens with a model projected every five years in the future. Without a multiplier you have to adjust the consumption for each of the 24 hours of the day for each calculation. That is to say, if the consumption increases by 38.5% you will have to multiply the consumption of each hour of the day by 1.385. If you have used multipliers, you leave the 24 multipliers as they are and simply adjust the average consumption by multiplying it by 1.385, saving 23 multiplications.

I have been talking about using the average consumption. In reality you can use multipliers with respect to any number. If you were using multipliers to define a pump pumping at different hours of the day, then it would be simpler to use the pumping flow (for example 8 l/s) when it is in use instead of average flow for the day. If it was running the first eight hours of the day, multipliers and flows would be:

Multipliers	1	1	1	1	1	1	1	1	0	0	0	0	0	0	0	0	0	0	0	0	0	0	0	0
Flow in l/s	8	8	8	8	8	8	8	8	0	0	0	0	0	0	0	0	0	0	0	0	0	0	0	0

This is a lot more visual as binary, 1 is "on" and 0 is "off".

The daily consumption pattern

You have calculated what the users consume in a whole day but the way the water is consumed in that day is as important as the total quantity. The figure below shows the variation of consumption according to the hour of the day for twenty urban users in Bolivia. Although each one consumes in a different way, tendencies can be observed. For example, the consumption at night, from 0:00 to 6:00 is very low.

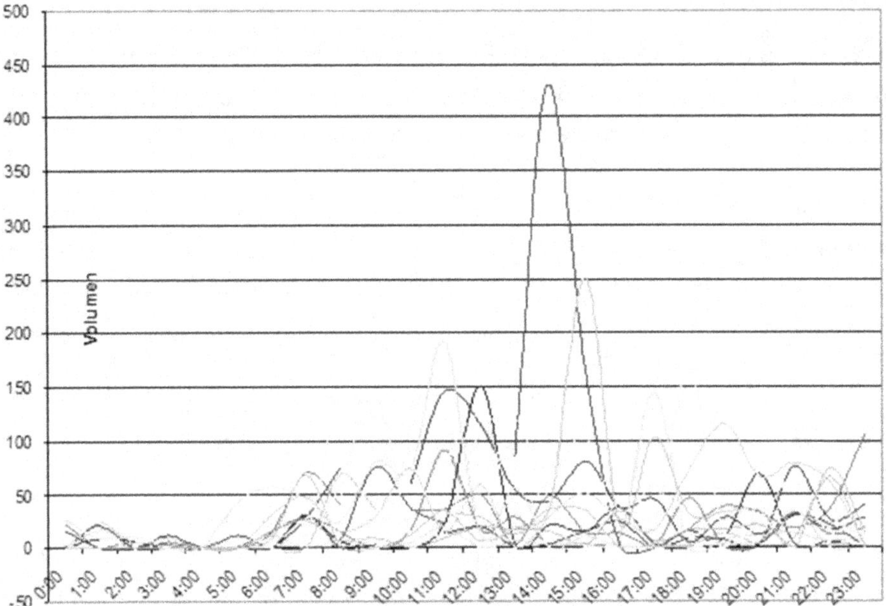

These tendencies can be summarized in the **consumption pattern** that you will find on the next page. The method of construction is relatively simple:

1. Take the measurement during 24 hours in certain points. In order to have statistical validity there have to be at least 30 points.

2. Find the average for each hourly frame, this sometimes requires the decision to be taken as to whether an unusual consumption is deleted or not. It's possible, for example, that there is one in every 30 consumers that is a high user therefore gave a peak that is not due to an error (i.e. a tea shop or car wash). Before eliminating them you should think hard about the decision. If you want to split hairs, you can see what values are anomalous in a statistical test of out of range values (i.e. Grubbs).

3. These tendencies are now reflected in what EPANET calls "patterns" but we have been calling consumption pattern. The dotted line is the average consumption.

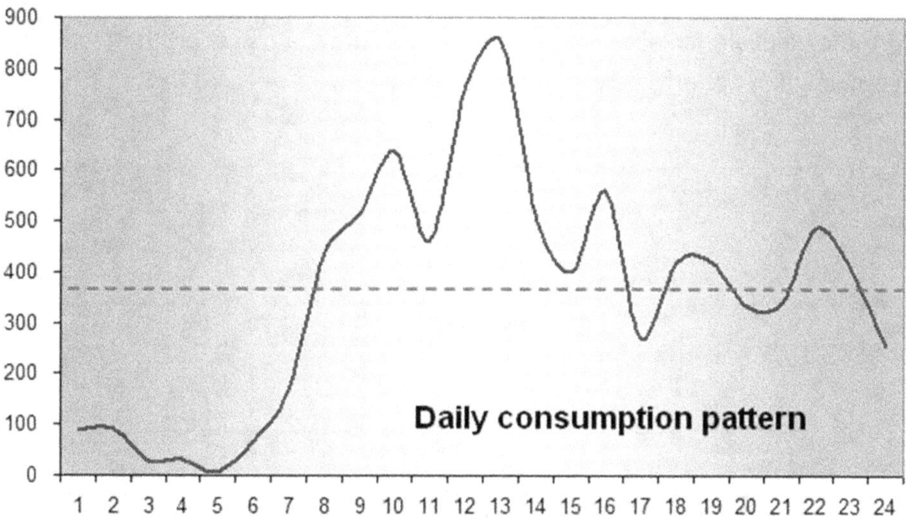

Note that if you measure patterns of networks that are not performing correctly you can end up with an undesirable pattern.

So that EPANET can understand the tendencies, they must be expressed in the form of multipliers, one every hour timeframe. In the following section you are going to see a step-by-step example.

If the multipliers are done well, their average is 1 and their sum 24. Once ready, enter them in EPANET by going to the browser and selecting *Patterns* in the *Data* tab.

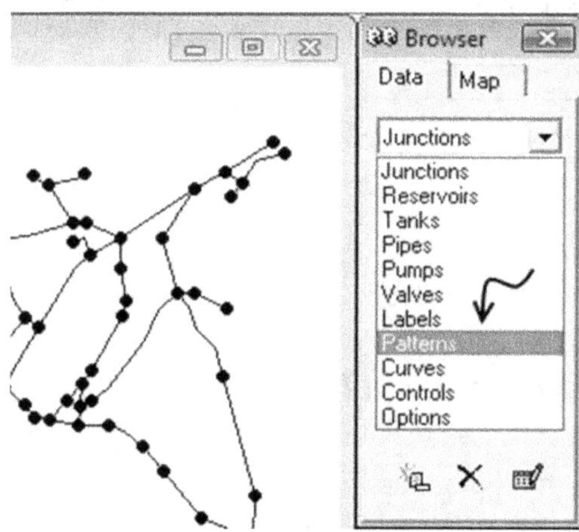

To open the dialog box, click the *add* button.   

By adding the multipliers for each hour, it is going to draw a bar chart that corresponds to the daily consumption pattern, as you can see below:

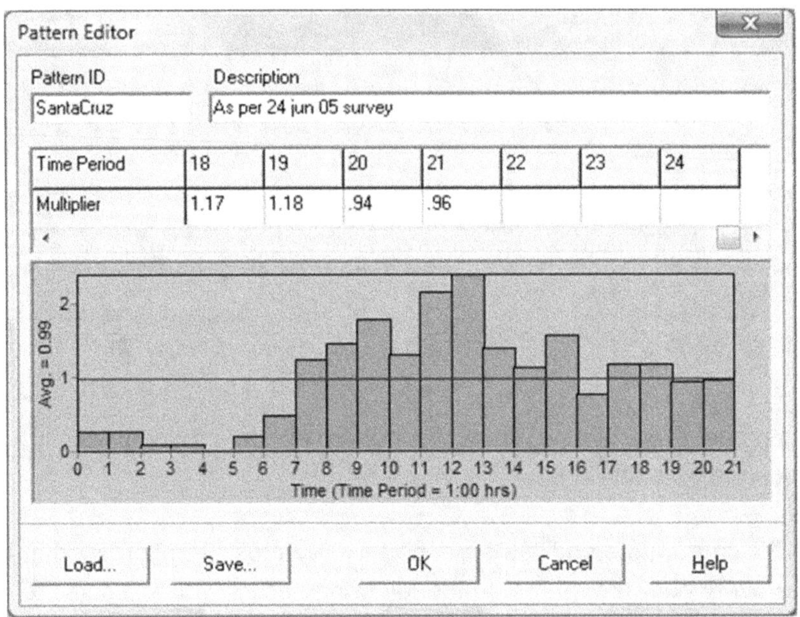

To name this modulation curve or pattern "1", enter "1" in the property *Pattern ID* for the junctions that have this consumption pattern. The highest multiplier is that **daily coefficient** and is of great importance in the calculation. In the example, although it is hidden it is 2.39.

Check out my channel to see future videos on the subject

 tiny.cc/arnalich

Example of the calculation of a daily consumption pattern

After having taken measurements in the field, the registered consumptions per hourly timeframe are shown in the first column. The next steps are:

	Hourly consumpt.	Multiplier
0:00	1800	0.20
1:00	700	0.08
2:00	200	0.02
3:00	300	0.03
4:00	500	0.06
5:00	1200	0.13
6:00	3000	0.34
7:00	8000	0.90
8:00	15000	1.68
9:00	12000	1.35
10:00	6000	0.67
11:00	5000	0.56
12:00	16000	1.80
13:00	23000	2.58
14:00	32000	3.59
15:00	25000	2.81
16:00	11000	1.24
17:00	7000	0.79
18:00	8000	0.90
19:00	9000	1.01
20:00	10000	1.12
21:00	9000	1.01
22:00	7000	0.79
23:00	3000	0.34
TOTAL	213700	24
Average hourly consumpt.		8904

1. Work out the total consumption for the day adding up the consumptions for each hourly timeframe:

$$1800 + 700 + 200 + \ldots + 3000 = 213,700 \text{ l}$$

2. Work out the average hourly consumption, dividing the total by 24 hours:

$$213,700 \text{ l} / 24 \text{ hours} = 8904 \text{ l/h}$$

3. Finally work out the multiplier for each hour by dividing the consumption for the hour in question by the hourly consumption:

$$0:00 \rightarrow 1800 / 8904 = 0.20$$
$$1:00 \rightarrow 700 / 8904 = 0.08$$
$$2:00 \rightarrow 200 / 8904 = 0.02$$

$$\ldots\ldots \qquad \ldots\ldots \qquad \ldots\ldots$$
$$\ldots\ldots \qquad \ldots\ldots \qquad \ldots\ldots$$
$$23:00 \rightarrow 3000 / 8904 = 0.23$$

4. To finish, make sure there isn't a mistake by checking that the sum of the multipliers is 24:

$$0.2 + 0.08 + \ldots + 0.34 = 24$$

The daily coefficient is 3.59.

When there is no data

Sometimes it's not possible to obtain data from the field, because there aren't counters or simply because there isn't a water system that allows the measurement of consumption. There are three ways to proceed:

a. Assuming that the population fits **a generic pattern**. Something that resembles the boa eating an elephant from *The Little Prince* will not be far from reality:

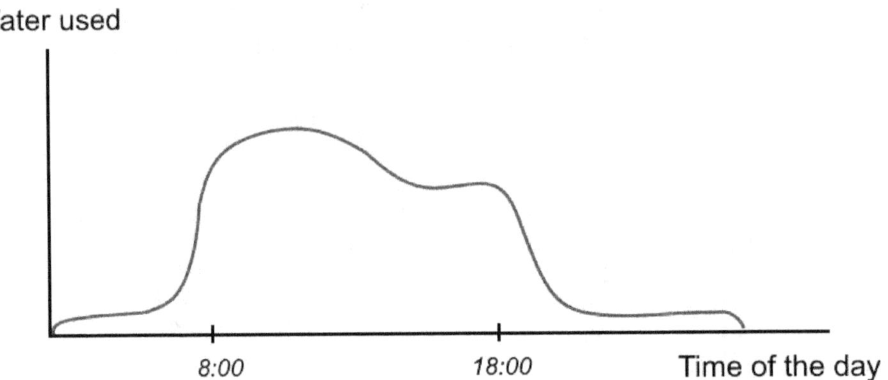

Low-income contexts tend to favor a higher consumption in the morning than this image, with a more accentuated valley at midday. In any case, the exact shape of the curve is more important for sizing a water tank than for sizing the network.

b. Assuming **a generic total multiplier** that summarizes the daily, weekly, monthly, etc., peaks around 3.5 to 4.5 for the overall coefficient, and 2 to 3 for the daily one (this will be explained in the following sections).

c. **Surveying the population.** A simple technique and very visual is for them to make piles with 100 grains of maize according to hours. If for the afternoon there are 40 grains, they consume 40% of the water. Unfortunately, people are not very conscious of how they use water.

Entering different demands

Take into account that **there isn't just one consumption pattern.** Taken to the extreme, each beneficiary has their own pattern. In practice, with a few patterns all significant groups are incorporated.

For example, compare an office's consumption, mostly in office hours,

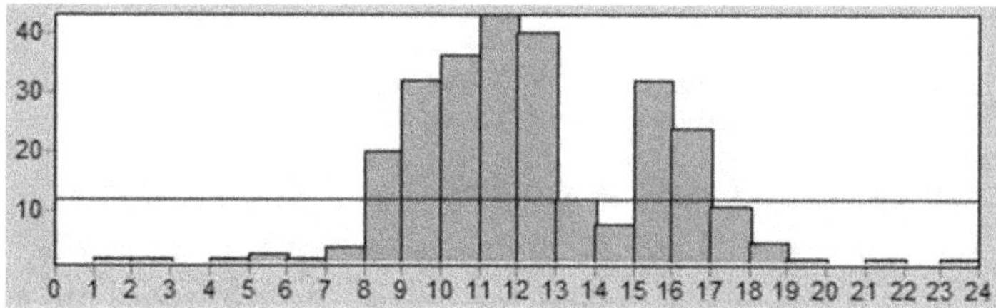

with that of a restaurant, that has peaks that coincide with mealtimes,

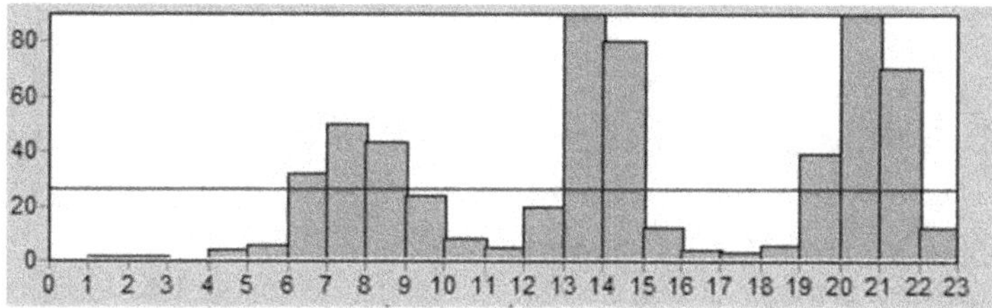

and to finish, with one from an industry that works at night because of excessive heat:

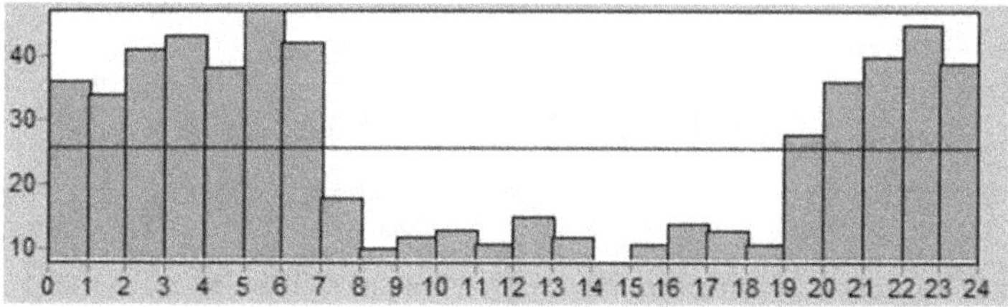

One junction can supply various families with a consumer demand category, and a workshop with an industrial demand category. This is done with the parameter *Demand Categories* in the properties of the junction, which opens the following dialog box:

The demand for this junction is the sum in each hour of the consumer demand category of 0.35 that will be applied according to the pattern 1 and the industrial category, 1.2 l/s, which will be applied according to pattern 2.

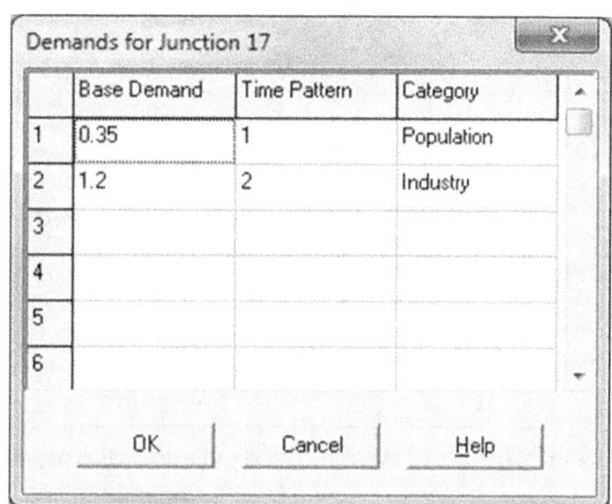

Demands for Junction 17

	Base Demand	Time Pattern	Category
1	0.35	1	Population
2	1.2	2	Industry
3			
4			
5			
6			

OK   Cancel   Help

The weekly consumption pattern

This is similar to the daily consumption, but with the following exceptions:

- It is unnecessary to build the entire pattern but simply compare the pattern for the measured day against the day of highest consumption in the week. To transfer the information use a coefficient called the **weekly coefficient.** If a Saturday was measured, whose daily coefficient is 1.18 and the day of highest consumption is Wednesday (daily coefficient 1.35) the weekly coefficient will be worked out as follows:

$$C_w = \text{highest multiplier / multiplier of the day of the measurement}$$

For example: $C_w = 1.35 / 1.18 =$ **1.154**

- Taking measurements over the course of a week is exhausting and normally there are not very significant differences. Therefore unless the population have a pronounced pattern (i.e. a market day) it is fruitless to gather data and contributes little to the result.

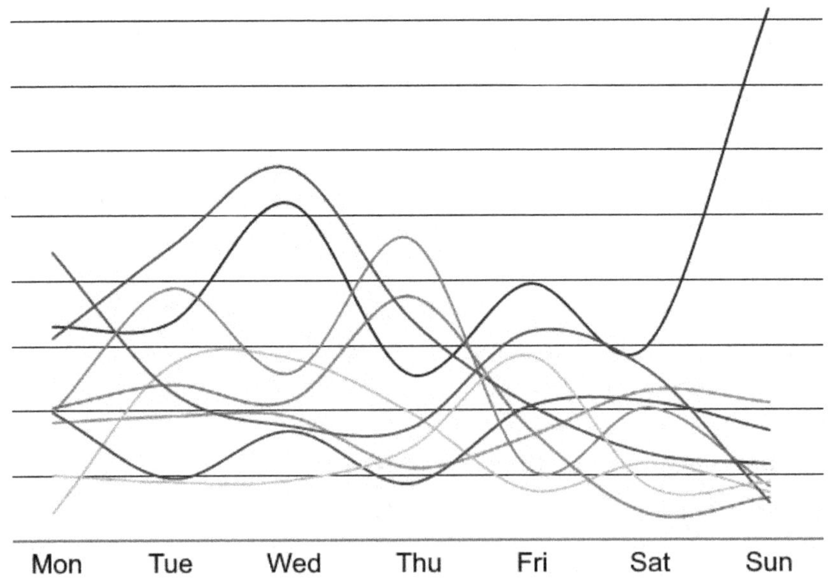

The value of the weekly multiplier is not entered into EPANET.

The monthly consumption pattern

This pattern is very important, and fortunately, as billing is normally done monthly, there invariably will be networks to study. We will proceed the same way as for weekly coefficient using this example below:

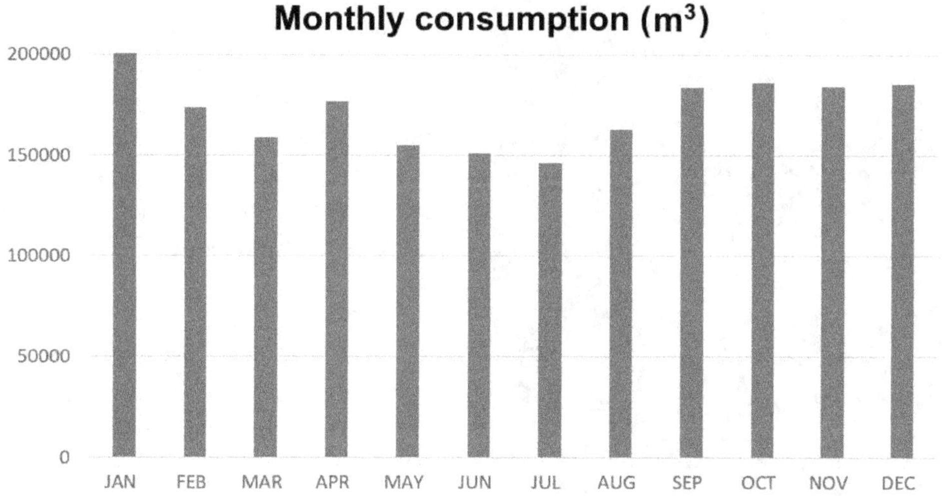

	m3	Multipl.
JAN	200282	1.17
FEB	173683	1.01
MAR	158623	0.92
APR	176411	1.03
MAY	155099	0.90
JUN	150940	0.88
JUL	146069	0.85
AUG	162857	0.95
SEP	183607	1.07
OCT	185982	1.08
NOV	183865	1.07
DEC	185429	1.08
Average	171904	

If the month of the measurement had been July with a multiplier of 0.85 and the maximum of 1.17 corresponded with January, the monthly coefficient is:

$$C_m = 1.17 / 0.85 = 1.376$$

The value of the monthly coefficient should not be entered into EPANET.

Unaccounted-for water

This refers to water lost by leaks, illegal connections, some public services, etc.

To make an approximation of this, compare the water that is produced against what is measured. The difference is the unaccounted-for water. For new networks it is estimated at 20%. Using this, the multiplier for a new network would be 1.2. So if the users are consuming 10 l/s, the network would really have to carry 12 l/s, and would lose 2 l/s along the way:

$$10 * 1.2 = 12$$

If the network is homogenous it is simple. Distribute the Unaccounted-for water between all of the junctions equally. When parts of the network are older than the rest, the unaccounted-for water should be increased for older pipes and when pressure is higher in a specific zone as it will lose more water. In both cases, a different multiplier can be established for each of the zones.

The value of the unaccounted-for water should not be entered into EPANET.

Global coefficient

To take the daily demand measured to the hour of the day, the day of the week, the month of the year with higher consumption, multiply together all the coefficients according to the type of analysis to be done:

a. Steady-state analysis multiplies all the coefficients,

Global = Daily x Weekly x Monthly x Unaccounted-for water
        = 2.39 x 1.15 x 1.37 x 1.2 = **4.54**

If the average demand for the future population was 100 l / hour, the maximum the system would have to deal with is: 100 l x 4.54 = **454 liters / hour = 0.126 l/s**

This is the value that should be distributed between the junctions, as you will see in the section Demand allocation, to be able to enter the parameter base demand in each of them.

b. Extended-period analysis multiplies all the coefficients except for the daily. That is because the multipliers for the daily pattern are going to be applied to each hour, and the multiplier for each hour doesn't have to be the maximum.

$$\text{Global} = \text{Weekly} \times \text{Monthly} \times \text{Unaccounted-for water}$$
$$= 1.15 \times 1.37 \times 1.2 = \underline{\textbf{1.9}}$$

If the average demand was 100 l / hour, the respective demands would be:

```
0:00     100 l/hour * 0.2 * 1.9     = 38 l/hour
1:00     100 l/hour * 0.35 * 1.9    = 66.5 l/hour
...      ...           ...     ...   ....
12:00    100 l/hour * 2.39 * 1.9    = 454 l/hour    Peak of consumption!
...      ...           ...     ...   ....
23:00    100 l/hour * 0.4 * 1.9     = 76 l/hour
```

You may have noticed that in both cases the peak is the same, 454 l/hour.

There is no need to add a safety factor since networks calculated this way are usually a bit oversized[11].

Summarizing

Look at this very simple example to summarize what we have seen up to now:

1. The population to be covered, all normal consumers, is 10,000 people.

2. You have decided that it is reasonable to project the population over 15 years. Applying the formulas, the design population becomes 18,000 people.

---

[11] *Cohen, J (1993). New trends in distribution research. Dynamic calculation and monitoring. Water Supply Systems. State of the art and future trends p213-250.. Computational Mechanics Publications. Southampton*

3. The total demand is 18,000 people x 50 liters per person = 900,000 liters

4. The average demand expressed in liters per second is:

    900,000 liters * 1 day / 86,400 seconds = 10.4 liters/ second

5. You have worked out the weekly coefficient to be 1.1, the monthly 1.4 and the corresponding unaccounted-for water 1.2. The adjusted demand is:

    10.4 liters/ second * 1.1 * 1.4 * 1.2 = 19.25 l/s

6. If you have 10 junctions, and have opted to assign the demand homogenously (next section) it leaves:

    19.25 l/s / 10 junctions = 1.925 l/s*junction

7. You have constructed your consumption pattern and the highest multiplier of the day is 2, that is to say, during that hour of the day the consumption is double the average.

8. Now you can proceed in two different ways according to the type of analysis you want to run:

    8.1 If you want to run a steady-state of the rush hour, the demand to enter into the properties dialog box is the product of the adjusted average demand per junction and the highest multiplier of the day:

    1.925 l/s*junction * 2 = **3.85 l/s**

    8.2 If you want to run an extended analysis, the succession of static analyses over 24 hours and not just of the peak hour, the demand to enter in the dialog box is exactly the one you have obtained in point 6, **1.925 l/s,** as well as adding the demand pattern calculated in 7, called "1" and shown by the arrow.

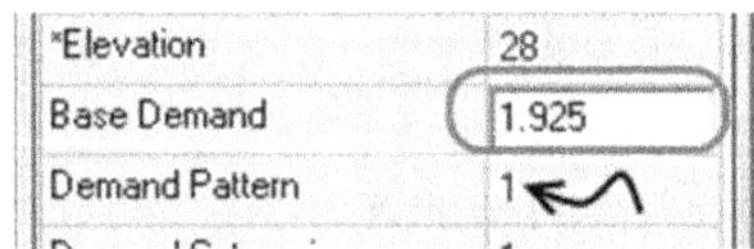

Approaches to calculating the design flow

Pay close attention to this section, as it´s really important. What we've seen up to now is the **time-varying demands method (TVD)** which is not suitable in all situations:

All taps open method

The TVD method is irrelevant in some cases where users don't choose how they use the system but are forced to line up for what the system can deliver, for example, in an emergency system set up after a disaster. It is also common where the population chooses a smaller less costly system, for example, a set of public fountains each of a certain capacity.

A tap is assigned with a specific number of users, and the flow is designed to supply all taps simultaneously. For example, one tap is assigned 250 people with a flow of 0.2 l/s. Then if two tap tapstands are used, the overall demand is 0.4 l/s and that is what is introduced in a node representing the tapstand in EPANET. There is no pattern.

The Simultaneity method

When a system is small using the average flow as in the TVD method will lead to installing pipes which are too small. To illustrate this, imagine for a second just one tap at the end of a pipe with a flow of 0.2 l/s. If at the end of the day the user has consumed 50 liters, the average flow is very small:

$$50l / 24h*3600 \text{ s/h} = 0.00058 \text{ l/s}$$

However, when the tap opens, the pipe has to carry 0.2 l/s...almost 350 times more!

This difference between the average flow and the instantaneous flow becomes smaller and smaller as the number of users increases. The fact that one user opens or shuts a tap becomes less and less important, and the average demand begins to stabilize. The point at which the simultaneity coefficient coincides with the TVD coefficient varies from one system to the other, but you can use the rule of thumb of 250 connections (not people).

This effect takes place in every pipe. Even if a system supplies a lot of people, if in one particular branch line there are only 35 connections, be careful not to install pipes which are too small.

The design flow of the pipe obtained by multiplying the average demand by a simultaneity coefficient you can get from this graph (Arizmendi 1991), that substitutes all the TVD coefficients:

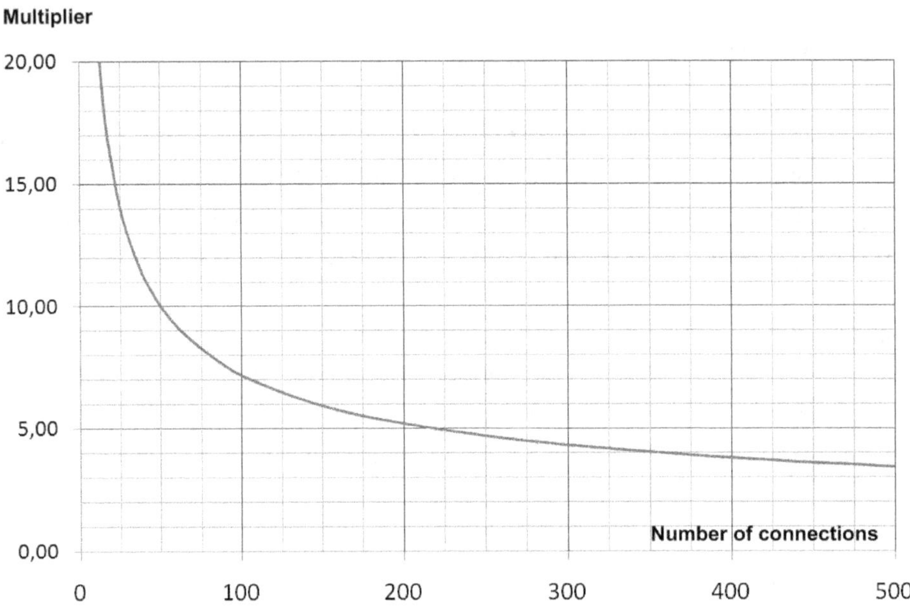

Now, there is an important catch, **you cannot use simultaneity with EPANET[12].**

The simplest way of avoiding problems is to establish a **minimum diameter** for the entire system (minimum 63 mm, better if bigger). Using minimum diameters is also really useful for protection against fires.

For large systems, there shouldn't be excuse for using pipes smaller than the minimum diameter, as you´re working with a bigger budget and the added expense is minimal. The concentration of more people also demands greater protection from fires.

To sum up, choosing the method for calculating the design demand comes down to 3 basic questions, which you can see in the flow diagram below:

---

[12] *Since each pipe will have a different number of users and a different coefficient, the mass is not conserved, in other words, water is "created or destroyed" when using this method.*

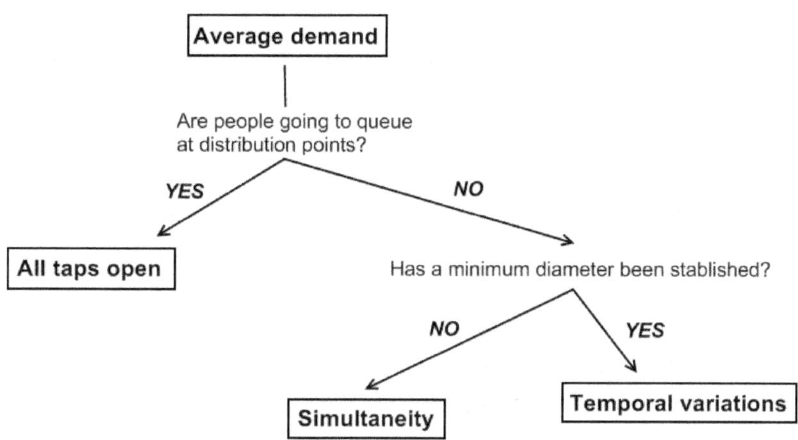

In all cases, **avoid installing pipes that are too small, especially over long distances**. For example, installing a 25 mm diameter pipe over a distance of 2 kilometers is a very bad idea. It will eventually clog up and it will be very difficult to find where. In addition, pipes between 12 and 63 mm are not very tolerant to variations in demand, or to changes in diameter due to limescale deposits or air pockets. Don't look too hard for savings in this area! It would soon end up costing the population a lot of money.

https://youtu.be/Jt6nGTZ5CgE   (Design flow)

Demand allocation

Previously we saw how to determine peak demand and how to apply it to projected population. Even more important is how to distribute this demand between different junctions, that is, to **allocate the demand.** I have a demand peak of 43 l/s, how do I distribute that demand between the 67 junctions that are on my drawing?

The demand per junction depends on how you draw the network, the number of nodes and how they are distributed spatially. **Allocating the demand correctly is one of the key steps in obtaining a precise model.**

You have these options and their combinations:

1. Point by point assignment

Assign each beneficiary their consumption. This option is laborious but makes models precise and is suitable for small networks or building interiors. It is impossible to predict consumption for future users therefore it is only suitable for networks with little potential for growth. Assigning point by point is especially recommended for large consumers, a hospital, a market, an industry…

To do this, distribute 50% of the total consumption between the ends of a pipe except in the case of a large consumer (the big yellow junction). Try to represent correctly the real distances to the end nodes in the model by precisely positioning the big consumer so that it divides the pipe into two, each with its corresponding distance to the end node.

The idea is that, since water will find the easiest way to the consumer, the different travelling distances to reach it from either side are accounted for.

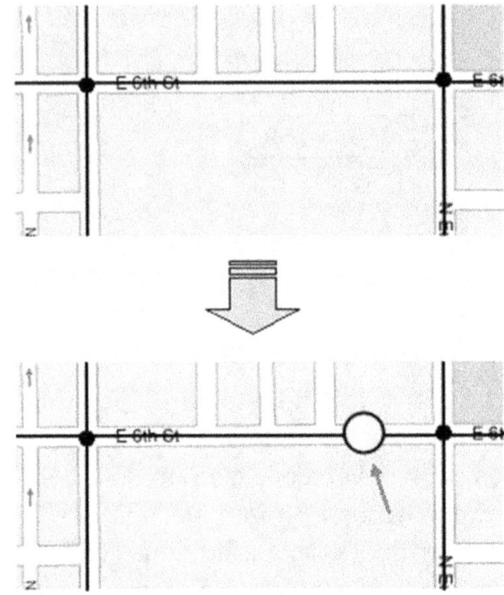

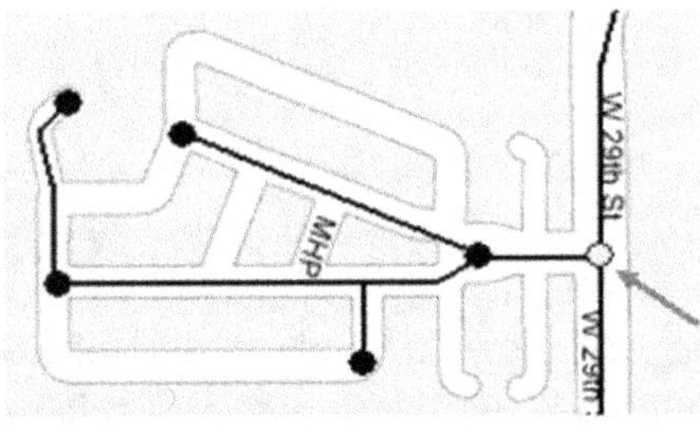

If there is a small branch coming off a junction, the total consumption of the branch is assigned to the junction (yellow).

2. Assignment by streets

The total users for each street are distributed between the first and last junctions. This can be done by stretch, as in the image, or by meters of pipes. For example, Silverlake Rd is 1200 m long, we are considering a stretch of 120 m and the total average consumption of the street is 20 l/s. The stretch considered will have a consumption of (20 l/s / 1200 m) * 120 m = 2 l/s, to be distributed between the two junctions.

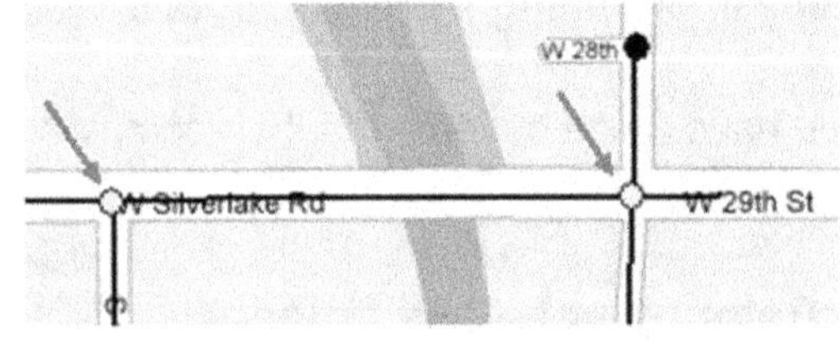

3. Assignment by grids

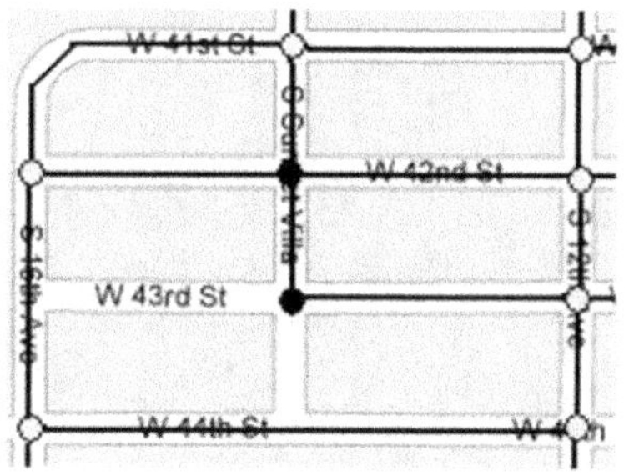

The consumption within a grid is distributed equally between the junctions surrounding it. This method of assigning is particularly useful if you use population densities as follows:

Maximum density: 500 per./km^2
Area enclosed by the grid: 2 km^2
Number of junctions: 10
Average consumption per person: 0.01 l/s

500 people/m^2 x 2 km^2 x 0.01 l/s / 10 junctions = 1 l/s in each junction.

4. Total assignment

Where consumers are evenly distributed in symmetrical networks the total demand can be divided equally between the junctions.

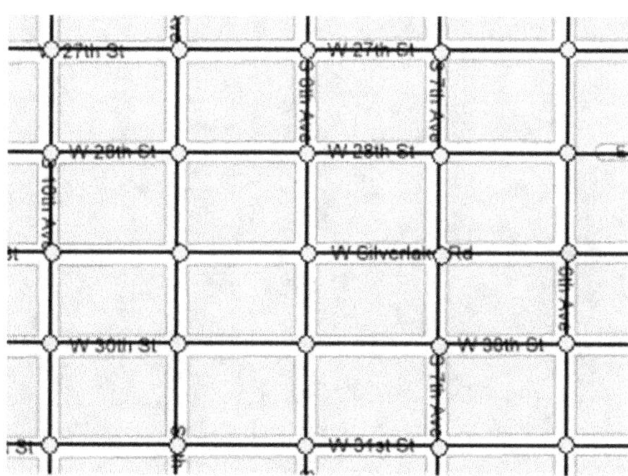

With a total consumption of 50 l/s between 25 junctions, the consumption per junction is:

50 l/s / 25 junctions = 2 l/s*junction

Modeling water quality

Introduction

For chlorine to have an effect, people must <u>drink it</u>. Of course chlorine doesn't work like a medicine, but…

    a. **Often people prefer to drink water that isn't chlorinated.** To build an amazing water system where people won't drink the water because it tastes of chlorine is failing to achieve the objective.

In some cases, the chlorination process is so bad that the users complain immediately that it is ruining their clothes. Often, communities unaccustomed to the flavor of chlorine, taste it and reject it even at low concentrations. A patient marketing campaign which introduces chlorine then gradually increases the dosage can deter people drinking from roadside puddles as illustrated.

For safe water to have an effect, it has to be drunk.

b. **It's essential to protect water from recontamination when** in contact with dirty containers, dirty hands, animals, etc. A small quantity called **residual chlorine** is needed, usually between **0.2-0.6 ppm**. Thus, if the person does not drink chlorine, one cannot be sure that the water has not been recontaminated. Take a look, for example, how water is recontaminated at the very collection point when some of the popular self-closing taps are used. They have a tendency to wash off the user's hand dirt into the container:

One last idea

Maybe hearing the word marketing applied to development has repulsed you. Don't underestimate the possibility of utilizing marketing to help introduce beneficial ideas, especially those that are culturally or religiously neutral. Examples and the philosophy behind the use of marketing are described briefly in *The Critical Villager* by Eric Dudley.

There are also interesting examples in developed countries. For example, to avoid saturating the environment with plastic bottles in Paris, the Water Authority launched the campaign "La carafe de Eau de Paris". It uses bottles like the one illustrated to promote the consumption of tap water in Paris, where 51% of the inhabitants say they drink bottled water.

Similar campaigns, adapted to the context, could promote the use of safe water instead of water from contaminated sources.

What parameters of water quality to model with EPANET?

Mainly two:

Water age

This is a measurement of how long the water stays inside the network. It has two main purposes:

1. **To ensure contact time with chlorine**. One of the conditions that ensure that the water is drinkable is that the chlorine has been in contact with the water for a certain period of time. A usual recommendation is 30 minutes, but the contact time depends on certain parameters, and can easily be double that. It is better to follow some actual calculation processes such as the *Log-4 virus inactivation*. To allow for this time, chlorination is usually done before the water tanks.

2. **To avoid quality deterioration with time.** When water spends a long time in pipes, the quality deteriorates noticeably. It you have come back to a house unoccupied for some time, you have probably noticed how dirty water comes out from tap. As a general rule, plan for **water to spend no more than one day in the network.** Although three days is normally recommended, it is possible that the network has not been well maintained and under these circumstances the risk is increased. Longer times suggest that the network has been overdesigned, or that it has a tree-like structure allowing the accumulation of water at the end of pipes. The image on the following page shows water age at 12:00 PM. It can be clearly seen that water has aged more at the far edges of the system.

From this we reach an important conclusion: **It is in the furthest and most isolated areas of a network where quality problems are greatest**, mainly because:

- The journey time is longer. Chlorine concentration decreases through internal reactions with time, increasing the possibility of contamination.

- Recirculation or dilution is not possible as the water only travels one way.

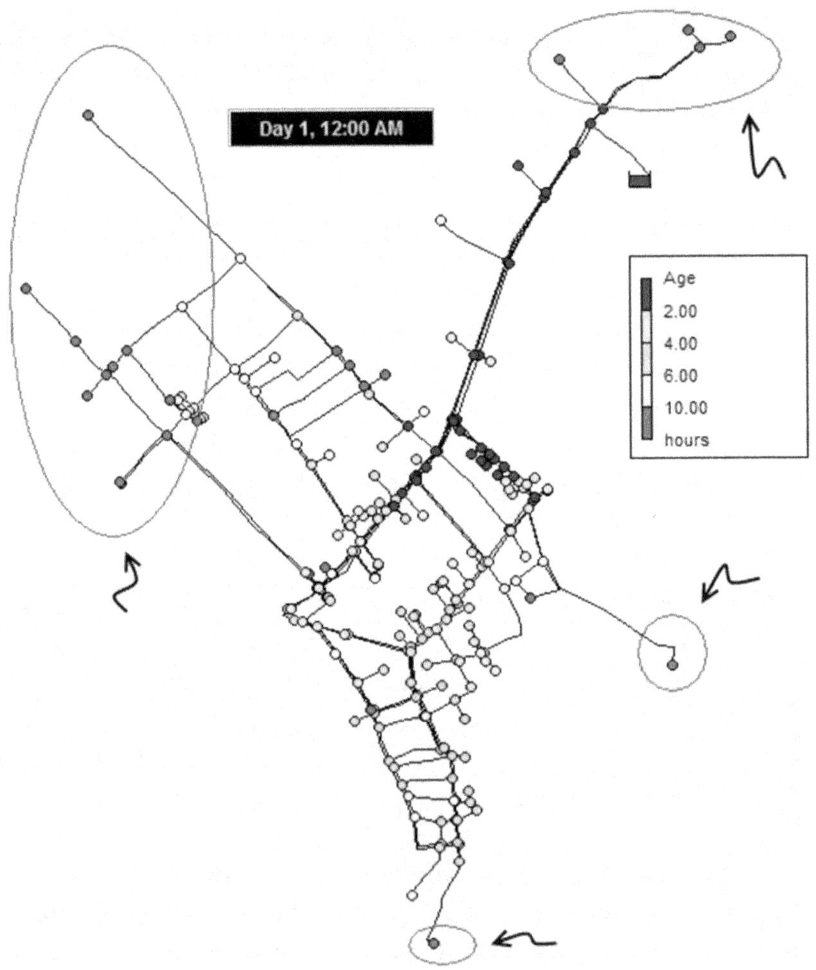

Chlorine concentration

We have already seen that to prevent water being contaminated after chlorination, it is necessary to maintain at least **of 0.2 ppm** of residual chlorine. When the amount of chlorine is increased, it starts to flavor the water and people may reject it in favor of less secure water sources. This maximum value depends greatly on the population treating the water and whether they are accustomed to chlorination or not. A **maximum of 0.6 ppm** is advised. Contamination is inevitable with opening taps by hand or using dirty containers.

Setting up EPANET for water quality analysis

You can only do all these analyses if you are simulating in extended-period, or quasi-static, as you have to see the change over time. You will find more details in the section Steady-state vs. Extended-period analysis in chapter 7.

To inform EPANET of your intention to work in extended-period mode, go to the Data tab of the Browser, and select Options. In that menu, select Times.

This choice opens the dialog box below to the side.
Set *Total Duration* to 72 hours. EPANET will show the evolution of parameters during that time. Only select 24 hours for very simple networks. For networks with tanks, using 72 hours allows you to observe cumulative effects from one day to another that would go unnoticed in a 24 hour analysis.

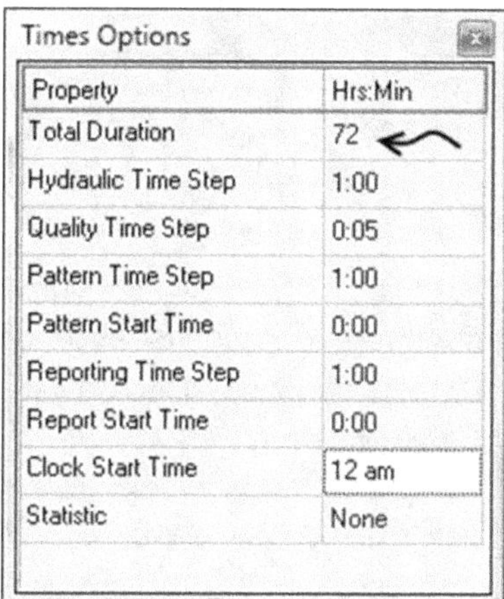

Property	Hrs:Min
Total Duration	72
Hydraulic Time Step	1:00
Quality Time Step	0:05
Pattern Time Step	1:00
Pattern Start Time	0:00
Reporting Time Step	1:00
Report Start Time	0:00
Clock Start Time	12 am
Statistic	None

In the following example, tank losses accumulate from day to day until, after 68 hours, it is completely empty. After this point, it is as if the tank does not exist, as it will never fill again.

It has two problems: it is probably too big (total height compared to the height of the second cycle) and it does not have enough recharge (does not recover between cycles).

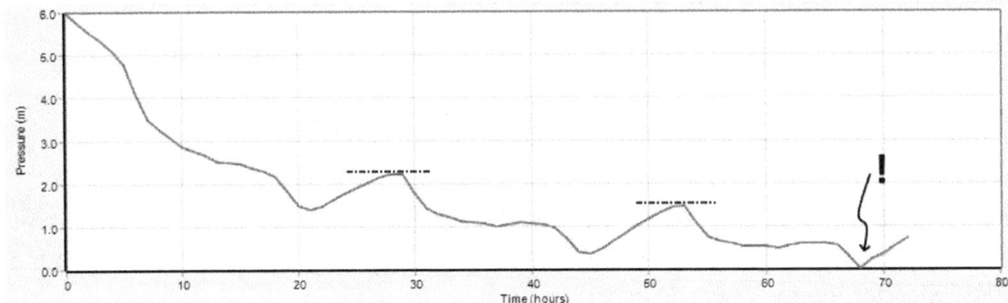

If the network is stable and properly designed, after a few days it starts cycling. Concentration, pressure levels and the tank level are very similar at the end of each day so that the network **goes back to its initial levels at the end of each cycle.**

The time that passes between each snapshot is called the **Time Step** and although it can be taken as milliseconds, months, or whatever, to make life simpler I recommend using hours. Human communities live according to the hours in a day, water age expressed in hours is easy to understand, and chlorine decay is sufficiently slow also to be evaluated in hours. To enter smaller intervals, for example a minute, will waste your time forcing you to go through the 1,440 images that the simulation will produce daily and more importantly, changes will pass unnoticed because they will be too subtle to be seen between if the step is too small.

The configuration of the first three parameters in the menu, Total Duration 72 hours, Hydraulic Time Step 1h and Quality Time Step 5 minutes looks like this:

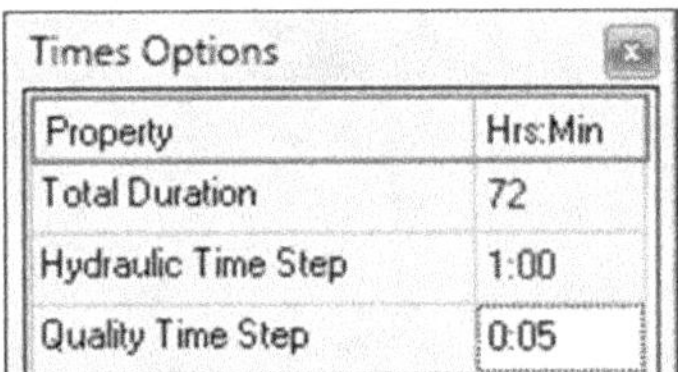

Times Options	
Property	Hrs:Min
Total Duration	72
Hydraulic Time Step	1:00
Quality Time Step	0:05

The remaining parameters allow more freedom for configuration and visualization of results but you will rarely use them except for *Report Start Time*, which we will look at later. Detailed explanations can be found in the EPANET User's Manual.

Configuring the quality

To set in EPANET the type of analysis to make, open the dialog box for quality in the Data tab, as shown below.

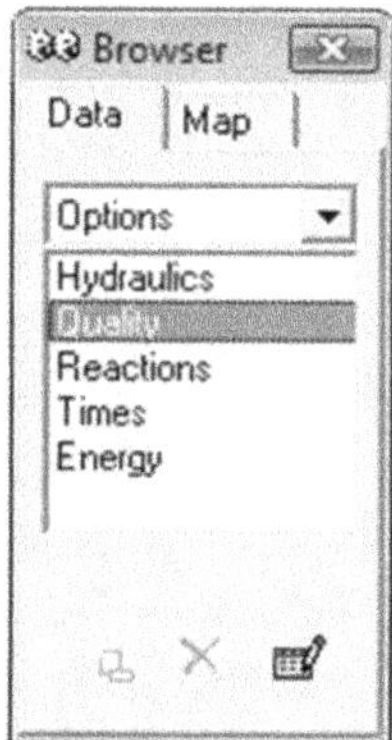

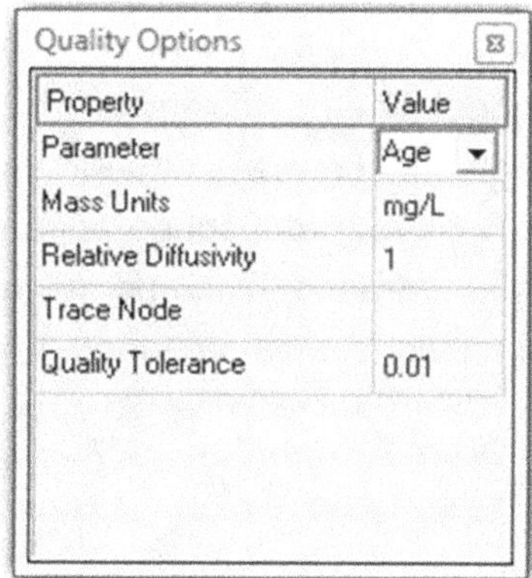

- **Chemical** is used to analyze the developing concentration of a reactive agent, usually chlorine.

- **Age** is used to see the water age.

- **Trace** is used to ascertain the percentage of water coming from a specific junction. This is useful where water is mixed from two sources and for example, when due to excessive salinity, it is necessary to dilute one source with the other to produce a mixed yield that is now acceptable.

To apply the configuration changes, press the lightning bolt.

Ascertaining the chlorine decay coefficients

These coefficients are specific to the water being used and the pipes carrying it. Therefore, you need to measure them, they can't be extracted from a book.

Chlorine is consumed in reactions in the water itself and through contact with the interior wall of the pipe. You can estimate this consumption through reaction coefficients.

Wall coefficient for chlorine decay

This coefficient is found experimentally and is elusive. The good news is that plastic pipes which are normally used are considered inert and have a coefficient of 0. If metal pipes are installed, wait to see what happens once the network is built, because you will have a "what came first the chicken or the egg?" problem. Should you order the material to suit the design, or should you design the network to suit the materials? You can use 0 in a design phase and measure it later if really needed. The coefficient can be measured by passing water of a known concentration at the constant flow rate through a long section of pipe, minimum 300 m. Use longer if the sensitivity of your measuring apparatus isn't good. By measuring the concentration of chlorine at the entrance, exit, and calculating chlorine consumption in the middle, you will see how much it consumes per length of pipe.

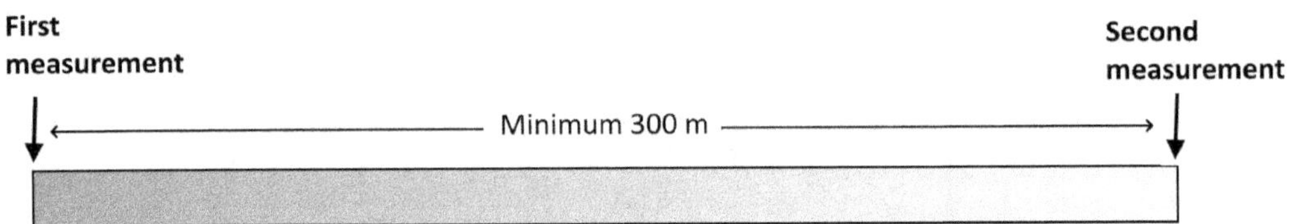

The pipes that you should suspect of high chlorine consumption are those of small diameter, metallic, and particularly if they are uncovered. The units are day^{-1}, negative values indicate that chlorine is consumed and positive values that it is generated. This value is entered in the pipe properties.

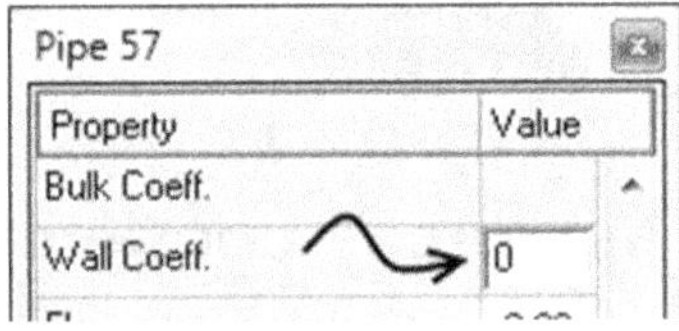

Bulk coefficient

This coefficient is determined experimentally, measuring how the chlorine coefficient in a glass container filled with water from the source changes over time.

To determine the concentration of chlorine use a pooltester which uses a pigment to dye the water according to the quantity of chlorine in the container then, when compared against the color scale tester gives clear visible results. Variations of 0.1 ppm are normal. Allow a sensible period of time between testing, for example, six hours to be able to compare color differences easily.

If concentrations in a working network are similar to what you are working with, subtle color changes will be difficult to measure with precision. If you measure higher concentrations, the speed of the reaction is different. To solve the problem test two samples side by side, one with half the concentration of the other. Remember that **one mg/l is the same as one ppm (part per million).**

Temperature can make the coefficient vary greatly, up to 15 times more at 5° then at 25°. Do the test at a temperature close to that found at the burial depth.

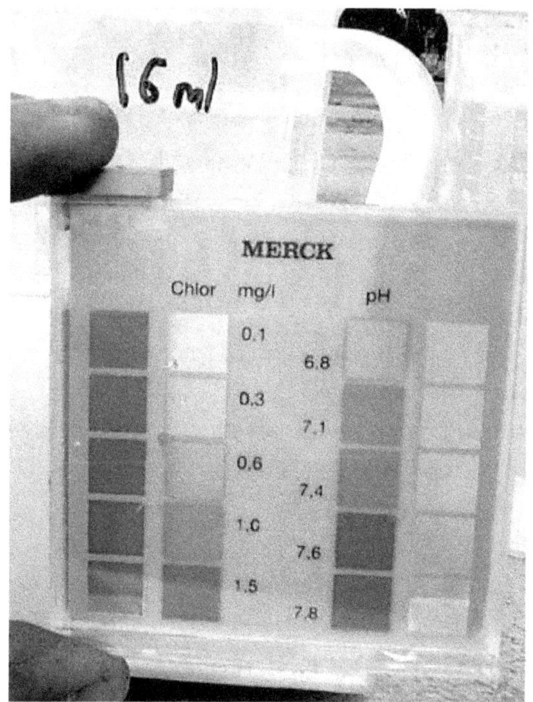

The procedure is as follows:

1. Add chlorine, for example from bleach, to the water to be tested until you measure a chlorine concentration of 1.5-2 mg/l (1.5-2ppm) with the pooltester. If you use bleach try with 4 to 5 drops. This will be the first sample.

2. For the second sample either halve the quantity used for the first, or just add to a fixed volume of the first solution the same volume of water. The idea is that the second solution ends up with half the concentration of the first.

3. Write down the concentration measured when the two samples are prepared. These are the initial concentrations.

4. From this point, measure every 4-6 hours until the concentration has dropped in the first sample to the last mark above 0 on the pooltester.

5. At this level you can ascertain the coefficient by apply the following formula:

$$K = \frac{\ln \frac{C_n}{C_0}}{t}$$

Where: K, bulk coefficient in days^{-1}
Co, the initial concentration
Cn, the concentration at the time of the measurement n
t, time in days

6.   Find the average of the two samples and use this value.

For example, if the initial value was 1.2 and you measured 0.6 after 48 hours,

$$K = \frac{\ln \frac{0.6}{1.2}}{2} = -0.3465$$

Defining the chlorine entry points

In low and middle-income countries' networks chlorination frequently happens in tanks. This guarantees that the contact time is sufficient for the chlorine to take effect. Two other possibilities are for a chlorinator which adapts to the circulating water flow or one with a fixed quantity feed. The main problem with fixed chlorinators is that the final chlorine concentration is very variable. Imagine for instance a fixed drip chlorinator. In flowing water chlorine is washed away maintaining stable chlorine levels. If the flow slows down the chlorine level will build up due to lack of flow.

Initial quality vs. Source quality

Both parameters are used to determine the concentration of all water that enters in a particular point whether it's a reservoir or a junction:

- **Initial quality**, is a parameter created to avoid computer calculation time by allowing it to start with situations closer to equilibrium. When dealing with reservoirs it can cause confusion as not only is it the initial concentration but also the concentration of the water entering the network. That is to say, if the reservoir has a concentration of 0.6ppm of chlorine, all the water entering the network during the simulation will have the same concentration.

- **Source quality** is used when you want to vary the *Initial quality* with a modulation curve. This is a similar situation to demand, and as we have seen with that, you should enter a value and afterwards a modulation curve that contains the multipliers for each time frame.

For example, to simulate a chlorinator that works for the first 8 hours in the morning with a concentration of 1ppm, the modulation curve that we would create (called "on-off") and the dialog box that appears when you click on the suspended dots in the reservoir properties is as shown below:

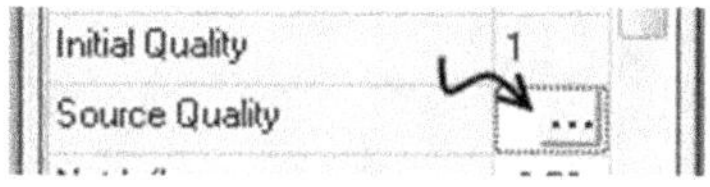

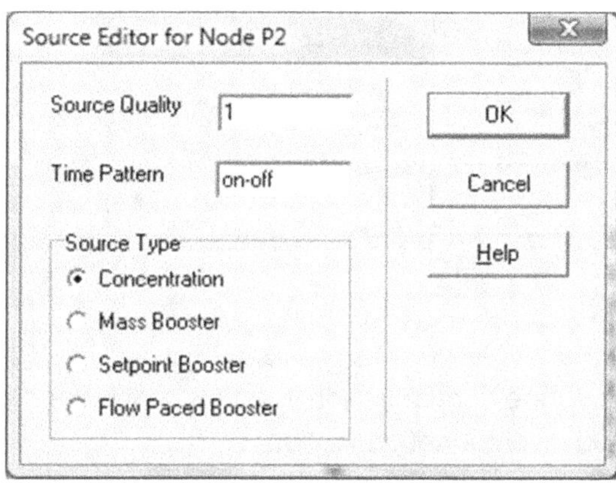

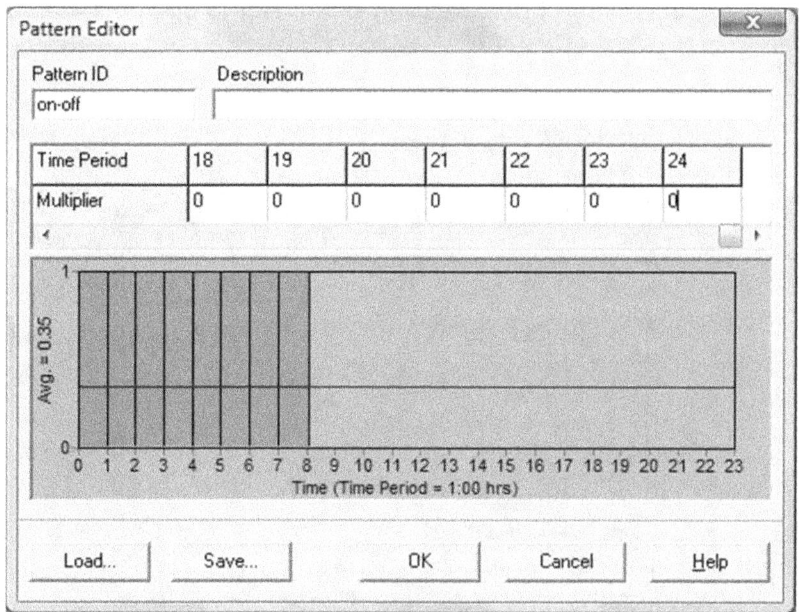

Chlorination of a borehole

This is one of the simplest cases and usually chlorination happens with a fixed chlorinator that switches on and off with the pump. As pump flow is fairly constant it is not necessary to have much regulation. The chlorinators are more robust, cheaper, and their start/stop mechanism is simple. Boreholes can be represented in two ways in EPANET, as a junction with a negative demand or as a reservoir. In both cases to incorporate the chlorinator, assign an initial quality to the junction or reservoir, as shown in the following images.

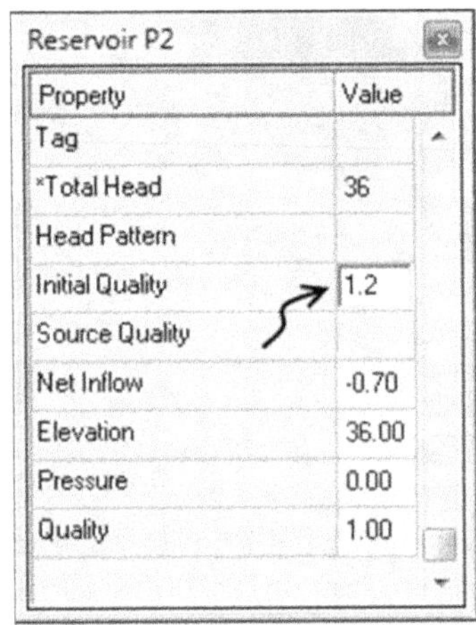

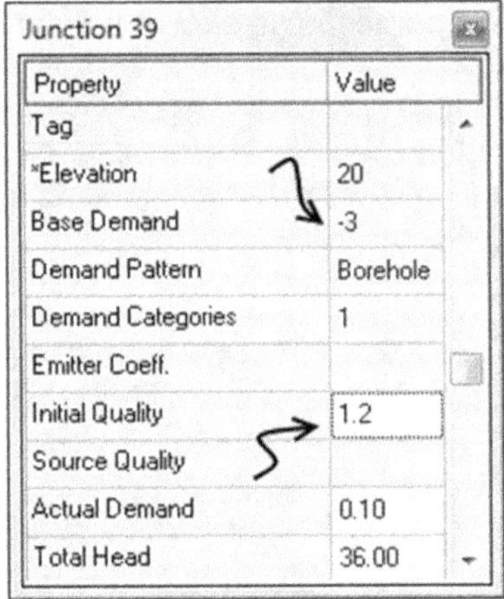

Satellite chlorinators

These are used to increase the chlorine concentration in points that have dropped below the desired level. They are also called re-chlorination stations.
Click on the junction or tank where you want to situate it to bring up the properties editor.
Click on the suspended dots of the *Source Quality* parameter to open the following dialog box:

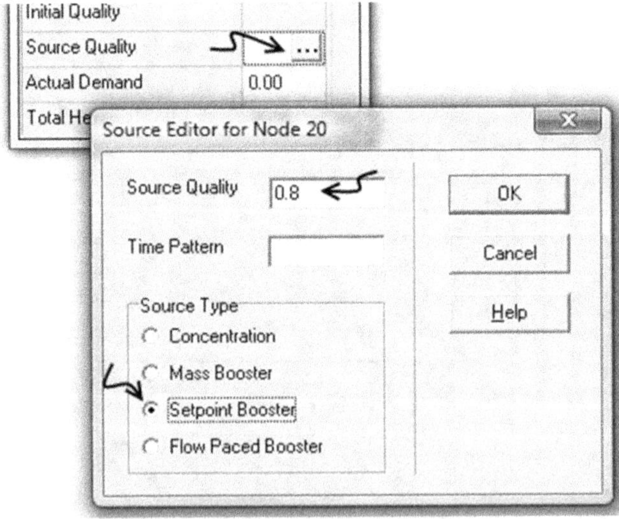

If the chlorinator is increasing the concentration in a certain quantity, select *Flow Paced Booster* and enter the value it adds. This is the most commonly used. If the objective of the satellite chlorinator is to restore a certain concentration value to the water, choose *Setpoint Booster* and then the value, which in the image is 0.8ppm. This applies for treatment stations, for shock treatment, followed by an activated carbon filter that eliminates chlorine and a subsequent re-chlorination to the desired value.

Full tank chlorination

For some simple systems every day the tank is filled, a quantity of chlorine added, and then is emptied completely. Here there are two ways to model this:

a. Providing you have modeled tanks as tanks, enter in properties the average bulk coefficient as the *Reaction Coeff.* then after adding the chlorine in *Source Quality* enter the chlorine coefficient. For a bulk coefficient of -0,4 day^{-1} and a concentration of 0.8ppm it would look like this:

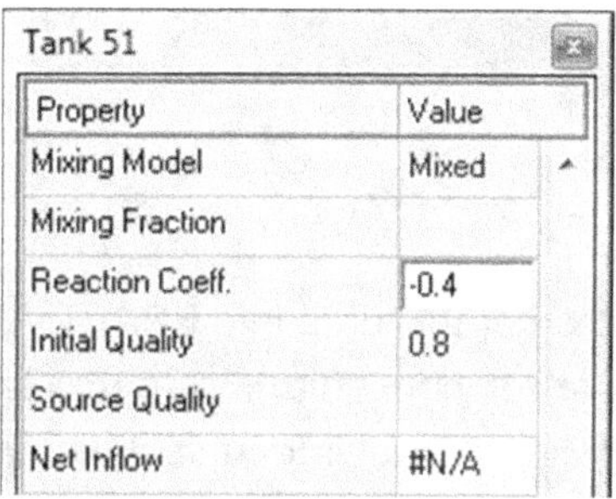

b. If you have modeled as reservoirs, enter 1 as the initial quality and create a behavior curve that simulates the consumption of chlorine in the tank. To construct this curve use the average bulk coefficient equation again, with the concentration at time taken out,

$$C_n = C_0\, e^{kt}$$

K, average coefficient in days^{-1}
C_c, initial concentration
C_n, concentration at hour n
e = 2.7182…

If the initial concentration in the tank is 0.8ppm and the average bulk coefficient is -0.7 day^{-1} the points for this curve are plotted as follows:

Hour 1, $C_1 = 0.8 * e^{-0.7 * (1/24)} = 0.78$ ppm
Hour 2, $C_2 = 0.8 * e^{-0.7 * (2/24)} = 0.75$ ppm

….   …   …   …

Hour 24 $C_2 = 0.8 * e^{-0.7 * (24/24)} = 0.4$ ppm

The graph shown below is typical for chlorine consumption:

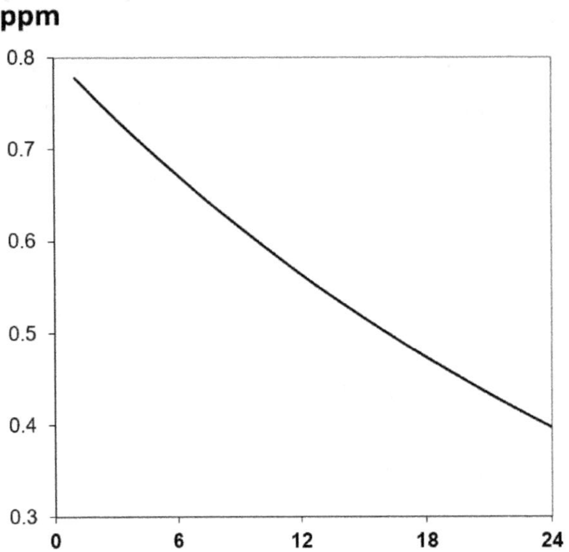

To enter this data into EPANET, select *Patterns* from the *Data* tab of the *Browser*. Enter the data for concentration (ppm) for each hour. Having 1 as the initial concentration avoids having to calculate multipliers, the values can be entered directly.

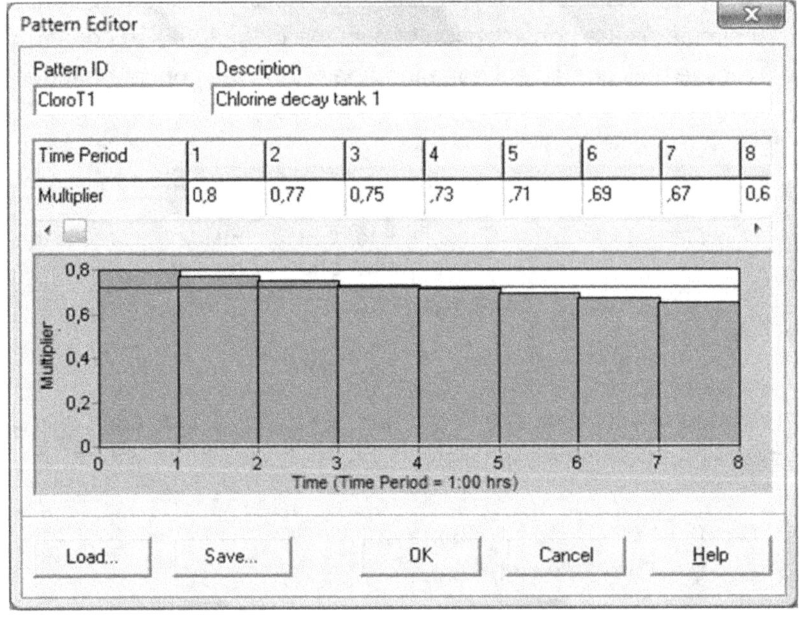

Time Period	1	2	3	4	5	6	7	8
Multiplier	0,8	0,77	0,75	,73	,71	,69	,67	0,6

CHAPTER 6

Analyzing the model

Beating the "Post-assembly laziness syndrome"

By this point you will have noticed that to set up a model requires a considerable amount of work. Sometimes time is short. Sometimes you have the false impression that you have done enough. Whatever the case, **a very common error is to rush the analysis.**

> *The harder I work, the more luck I seem to have.*
>
> *(Thomas Jefferson)*

Here it's important to have clear ideas. You have spent all that time compiling data so the computer can give you a good analysis and now you should take advantage of that. Use it to the best of your ability to compensate for all the time you have invested in assembling the model and you will be rewarded with conclusive results. If you're not prepared to take time analyzing the results and working for the best design possible then don't turn on the computer in the first place. Get rid of this book and look for one more entertaining!

Every minute you invest in the analysis will save hours of headaches, changes and delays in a network that doesn't work. Badly planned networks lose their objectives before they are even built, however, with a small amount of your time you will add so much value for beneficiary communities.

It is human nature and inevitable that we all experience this syndrome at some point. The best way to combat it is:

1. **Start soon**. As with anything, don't spend time worrying about possibilities but focus your attention on the immediate problem. Once you have started, particularly tasks that make you worry or feel lazy you will find it much easier to continue.

2. **Establish clear objectives for this phase.** "After the analysis I want three alternative designs under 80,000 €, or "I am going to find one solution that involves installing less that 3 km of pipes". A lot will depend on the characteristics of your job, but to set targets will keep you focused.

3. **Helping rushers to understand.** Analyzing a fairly complicated network can take 1 or 2 days. Do you really need to rush so much? Is it so important that so-and-so has a piece of paper on such-a-day to jeopardize the outcome of the entire project? By avoiding speculative guesswork and presumptions you will undertake the job properly. If you have managed your time badly or time is limited, this is not the place to save time, sacrifice other things.

Steady-state vs. Extended-period analysis

Steady-state only evaluates a single instant in time, generally the most unfavorable for the network. That is, it's like freezing time in that worst instant to see the velocities in the pipes, the pressure at the points, etc. As an analogy we can say that steady-state is the moment of the total solar eclipse:

Extended-period analysis (quasi-static) evaluates a succession of instants, usually an image of what happens each hour in a day, not just one of them, and is normally shown on the screen like a film. This is also called quasi-static. Going back to the analogy of the eclipse:

To start modeling the design with Steady-state makes it easy to make changes and see the results quickly. In this phase you can check if the capacity of the network is enough to meet demand using the calculation criteria as explained in the next section. Once the model is stable and effective for the demand we move to extended-period analysis to look at the parameters that depend on time, like the level of the tanks or the chlorine concentration.

Viewing steady-state and extended period analysis in one go

In practice, the easiest way is to carry out a spreading by starting to display the results at times of highest consumption. Therefore, without making changes or readjustments, you can go from a steady-state analysis (ignoring the rest of the time periods) to one in an extended period (running the results to see the rest of the time). To do it:

1. Find out what is the time of day of maximum consumption, for example, 13:00.

2. Go to >*Browser/Data/Options/Time*.

3. Enter the Total Duration, at least 72 hours, and the Results Start Time, 13:00.

After having configured the dialog box as in the image, EPANET will show you the evolution of the behavior of the network 24 hours a day starting with the image corresponding to the steady-state analysis, without the need to make any adjustments.

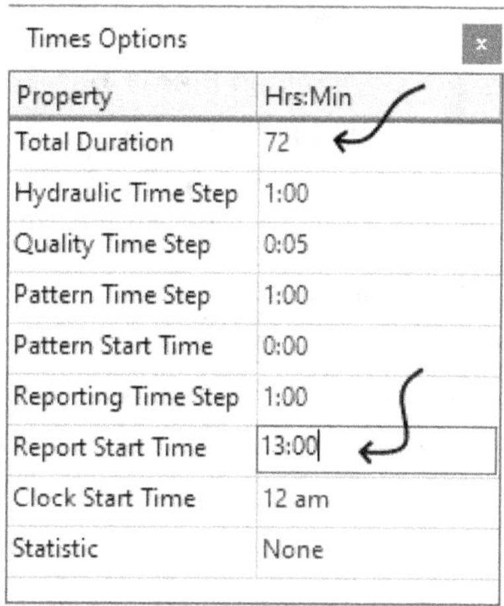

Times Options		x
Property	Hrs:Min	
Total Duration	72	
Hydraulic Time Step	1:00	
Quality Time Step	0:05	
Pattern Time Step	1:00	
Pattern Start Time	0:00	
Reporting Time Step	1:00	
Report Start Time	13:00	
Clock Start Time	12 am	
Statistic	None	

Calculation criteria

Before starting to analyze the model the calculation criteria need to be defined, that is to say, the range of values in which the solution for the model will be acceptable. Modeling in a development context, we are mainly interested in four parameters: pressure, velocity in the pipes, water age and chlorine concentration. We have already covered the last two in the previous chapter.

Pressure

Usually you would try to keep this **between 1 and 3 bars at the tap for every user**, or 10 and 30 meters of head, which is the same. This can be easy or difficult, depending a lot on the terrain.

Networks with less than 1 bar have problems. For example, very low pressures mean that self-closing taps don't shut. Another problem is that the remaining flow at low pressure is so small that the people get tired of holding them open, so they tie them as in the photo on the left. Higher pressures can cause leaks, make the system more prone to breakages and become inaccessible. One of the consequences is that children cannot open the auto-close taps. At more than three bars the flow from the tap can become a spray leaving containers half full when the bubbles disappear and soaking the people (photo on the right). Flooding problems are widespread.

https://youtu.be/Nub80L5KmOY　　　(Maximum and minimum design pressure)

Velocity

This is usually **between 0 and 2 m/s <u>at peak consumption</u>**. At very low velocities, suspended solids start to settle and accumulate in the lower areas, reducing the effective diameter of the pipes, but:

0 m/s!! A frequent and outdated recommendation is to guarantee a minimum speed of 0.5 m/s so that the network is self-cleaning. The latter is both absurd and dangerous because it leads:

- The construction of networks with too small diameters, which generate an increase in pumping costs and make future extensions difficult.

- It is impossible to put into practice. As the size population served changes, it is not possible to maintain such speed ranges for the life of the network even if you wanted to.

The networks can be cleaned thanks to the washouts at the lowest points that allow this cleaning, and the water must enter clean, not full of nuts and gravel as shown in this video (which makes the case for wastewater):

https://youtu.be/c1xX90ZfBj4   (Solids transport in waste water pipes)

Velocities more than 2 m/s suggest the pipes are too small and create a risk of damage by water hammer and tear.

Using Unit Headloss for "enlightment"

This is a measurement of energy that is lost by friction inside the pipe and depends primarily on the flow rate and its velocity. It is measured in meters of head lost/km of pipe. The more meters of head lost the more inefficient the pipe is. Accepted values are below **5 m/km for networks** and **0.04 m/meter in buildings**.

Occasionally it is desirable to have a lot of friction to dissipate pressure in low points in the network. This allows savings by laying a smaller pipe and prevents excess pressure providing you don't anticipate any future extensions to this line. Some other times it is too low, perhaps because of the relief or because an excessively large pipe was installed As you can see in the example in the image below, the thick red pipes (dark in B/W print) exceed this value.

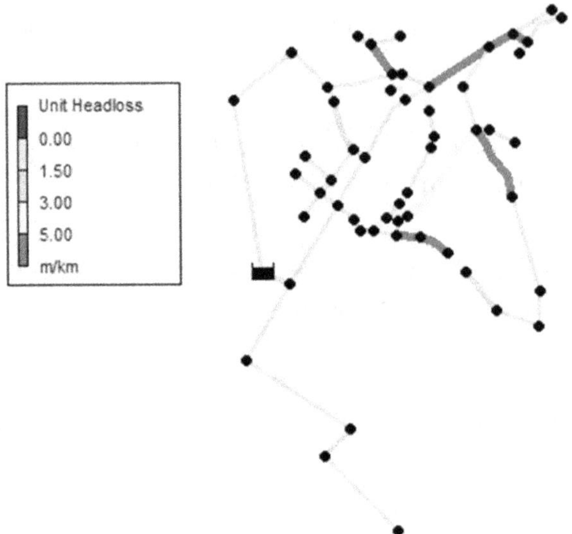

Thus put, it does not seem particularly interesting, but it is a vital ally to quickly understand what happens in complex networks and know where to start. This video explains it to you with examples:

https://youtu.be/LAB1Kuwv-t4   (Using head loss to diagnose a system)

The **head loss is not a calculation criterion, it is a tool** to understand your network. If the network asks for 30 m/km, perfect! And if it asks for 3 m/km, perfect too! There is no criterion to meet, it will depend fundamentally on the topography.

What to look for with what type of analysis

Now you know the basic calculation criteria and the two types of analysis, the following diagram will serve as a checklist, taking into account that every system is different and that there are situations where going outside the recommended values can be justified:

Steady-state:

- There are no pressures below 1 bar.
- There are no pressures above 3 bars.
- There are no velocities above 2 m/s.
- Investigate unit head losses above 10 m/km and below 2.5 m/km.

Extended period analysis:

- There are no pressures above 3 bars.
- There are no pressures below 1 bar in other hours.
- The chlorine concentration is not less than 0.2ppm.
- At taps, concentration is below 0.6ppm.
- The water doesn't spend more than 24h in the network.
- The water tanks refill with each cycle.
- The pumps are working within their ranges.

Without overwhelming you, these are the most useful and simple although there are many other checks you can do. I will suggest some courses of action you can take for each case in the next section.

A way of working

Now that you have built the model, you will be asking what you are going to do with it. Firstly, save a backup copy in your computer and have the discipline to only use it if you have destroyed the model you work on and need to start again. Afterwards, modify the elements so that they work as we have proposed. It is important to know that: **there are infinite solutions,** some better than others. Therefore, look for the cheapest, the easiest to build, consumes the least energy, etc. Generally, this means that you end up with various distinct solutions to compare (details in chapter 8).

1. Modify the base model by adding pipes, modifying its properties, changing its route until you arrive at a solution, say solution 1. Start again, you may want to try putting in a new tank, add a pump and modify pipe 54 until you arrive at solution 2. Continue like this until you find that you have run out of ideas and you have various solutions to compare.

2. Compare the solutions and select one of them.

3. Work with this solution again to look for improvements. Also, model what would happen if a pipe broke, if there was some very pronounced consumption in one zone, etc.

So far, you would work in static mode on, which is less cumbersome and allows you to hone in on a potential solution without hassle. From here you should already observe the rest of the screens of the analysis in extended period.

4. Check that the maximum pressures, water age and chlorine concentration are correct.

5. Lastly, review your model to look for mistakes. The checklist at the end of the chapter can help you with that.

6. Export to a spreadsheet the lengths and diameters of pipes used. Establish an approximate unit cost that includes the excavation, for example, 22 USD/m for a 90 mm pipe. Get the total to arrive at an approximate figure; it is not a detailed estimate, it is simply a figure accurate enough to be able to compare designs. If the networks have pumps, include the pumping costs.

7. **Benchmarking**. Start again by repeating cycle 1-6. Perhaps this time you will try putting in a tail tank, adding a pump and modifying the pipe 54 until you reach Solution 2. So on and so forth until you run out of ideas, you notice that you have a good understanding of how the network works and you can't get the number down. It's like the score of a video game and you're playing different games. If the network is very simple there may only be one direct solution, but as they get more complicated, they need more and more attempts. Besides cost, you can compare other parameters, for example, pressure.

Don't underestimate the importance of doing point 7. As an example, in a recent assignment to redesign 20 water networks with 425,000 users for UNICEF, I achieved average reductions of 26% in costs and peak flow increases of around 40%. Some of these networks required up to 9 attempts, and I have been calculating networks for 20 years! The result was a saving of approximately 5 million euros and 175,000 more people served, with only 182 hours of work. Few things you do in a development project will have that rate of return.

Correcting errors

Once the model is loaded it's time to calculate it. To do this, press lightning bolt. It will leave a message while it is processing:

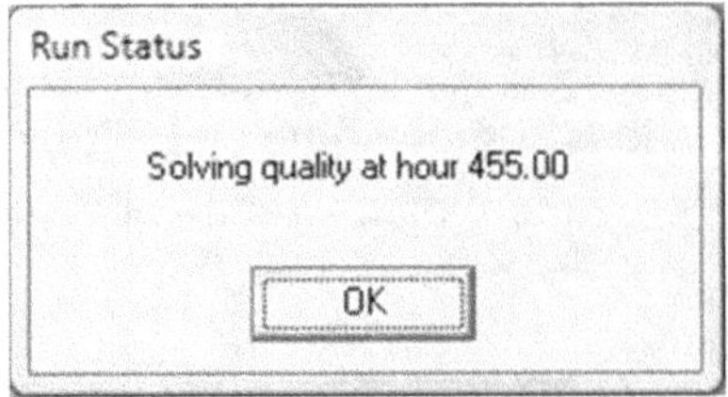

Afterwards, you will see an error message similar to this, except in the improbable case that you have a perfect network first time or your network is generously over-designed.

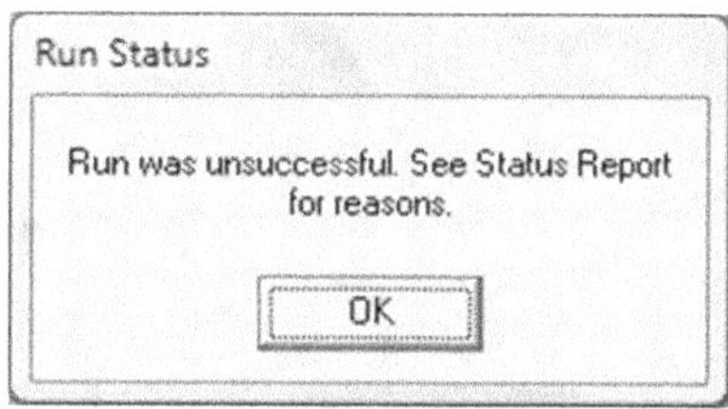

Don't worry, this is completely normal. You are entering a phase of eliminating possible errors. This phase is very simple and quickly resolved when you appreciate what's happening. We are going to see what these errors are.

Disconnected junctions

After hitting OK you will receive a message informing you of the errors. This message has some general content followed by the error messages, some more intelligible than others. In the case of disconnected junctions the message is very clear:

```
Input Error 233: Node 51 is unconnected.
```

Look for it in the general message as shown below:

```
Status Report

Page 1                                          Fri Dec 11 09:21:11 2009

****************************************************************************
*                            E P A N E T                                  *
*                    Hydraulic and Water Quality                          *
*                    Analysis for Pipe Networks                           *
*                         Version 2.00.12                                 *
****************************************************************************

Analysis begun Fri Dec 11 09:21:11 2009

   0:00:00: System ill-conditioned at node 4
   0:00:00: Reservoir 3 is closed

WARNING: Node 4 disconnected at 0:00:00 hrs
System Error 110: cannot solve network hydraulic equations.
Analysis ended Fri Dec 11 09:21:11 2009
```

This error is very simple to eliminate. It's saying that there is a junction or section of the network that is not connected. Normally it will be in one of the following two situations, as detailed on the enlarged networks drawings:

a)   You drew an extra junction without realizing.

b)   You had a bad aim clicking on one of the ends of a pipe and left an unconnected junction in the middle

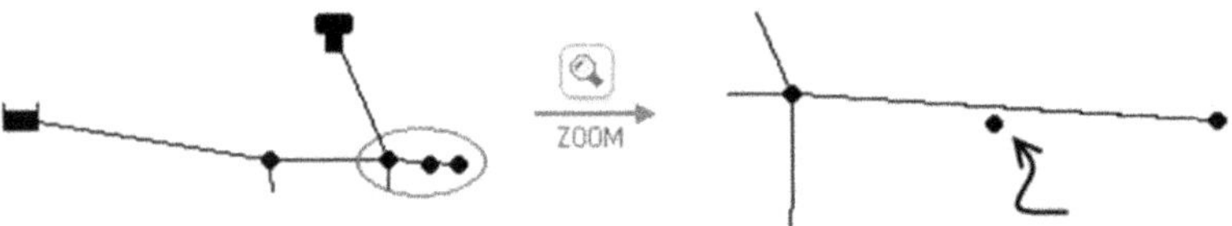

Pay particular attention to tanks as even although they might not be connected to the network, they do not show as an error. This means that to all intents and purposes they do not exist, but we will spend money building them because we think they are necessary. In the image the tank is not connected.

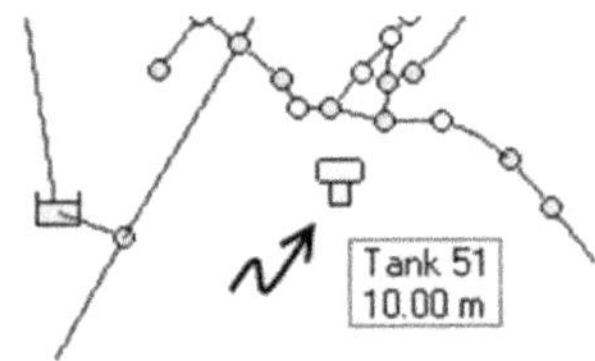

It can't solve the hydraulic equations

This error, that seems so intimidating, is actually quite simple:

```
System Error 110: cannot solve network hydraulic equations.
```

In the majority of situations, there is a part of the network that has a demand, but is not connected or there is unrealistic value somewhere, i.e., *Base demand*=10000000. It is similar to the previous error except it should have at least two junctions and a specified demand. When the two sections are connected, it disappears.

Sometimes this is the companion error message:

```
0:00:00: System ill-conditioned at node 57
```

Closed reservoirs

```
0:00:00: Reservoir P2 is closed
```

Basically, it's disconnected (without a pipe that connects it to the rest), as in the image. However it can also be the result of another error and disappears when this is resolved.

There aren't any tanks or reservoirs

```
Input Error 224: no tanks or reservoirs in network.
```

You need to have an entry point or a drain in the network for it to be able to be calculated. If you are missing one of these, even if there is a junction with negative demand (bringing water into the network), it needs one of these two elements.

You have not defined a certain parameter

It is obvious there will be many of these to do. Let's look at some examples:

```
Input Error 205: Junction 2 refers to undefined time pattern.
Input Error 206: Pump 2 refers to undefined curve.
Input Error 217: invalid Energy data for Pump 2.
```

Tag or junk errors

You will have noticed the end tag:

```
Input Error 200: one or more errors in input file.
```

This is a good example of many error messages that appear in the box that in reality don't refer to anything and can be ignored as they will disappear when the significant errors are resolved. For this reason, start to eliminate the errors that you know. If you resolve one of the known errors and leave the unknowns to the end, most probably they will disappear during the process of elimination.

To illustrate this point: notice that to inform you a pipe is missing between two parts of the network, EPANET gives you four different error messages. They all disappear when the two junctions are joined with the rest of the network.

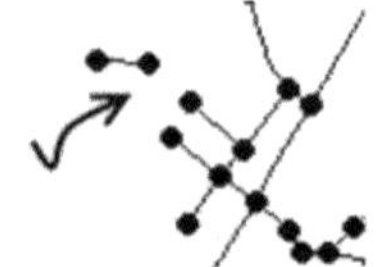

```
Status Report                                                  □  ▫  ✕

Page 1                                          Fri Dec 11 09:21:11 2009

  *********************************************************************
  *                         E P A N E T                              *
  *                  Hydraulic and Water Quality                     *
  *                  Analysis for Pipe Networks                      *
  *                      Version 2.00.12                             *
  *********************************************************************

  Analysis begun Fri Dec 11 09:21:11 2009

    0:00:00: System ill-conditioned at node 4
    0:00:00: Reservoir 3 is closed

  WARNING: Node 4 disconnected at 0:00:00 hrs
  System Error 110: cannot solve network hydraulic equations.
  Analysis ended Fri Dec 11 09:21:11 2009
```

Up to this point the most common errors have been described but if you encounter other errors, you can find more help in the EPANET User's Manual under *Error messages*. Remember that EPANET has a help section to assist you. Now we concentrate on the error you work with …

Negative pressures

```
WARNING: Negative pressures at 0:00:00 hrs.
```

In other words, there are points in the network where the water doesn't arrive. Modify the legend as explained in the section Basic tools. If the first point is set as 0, then these points will be in dark blue as indicated on the image:

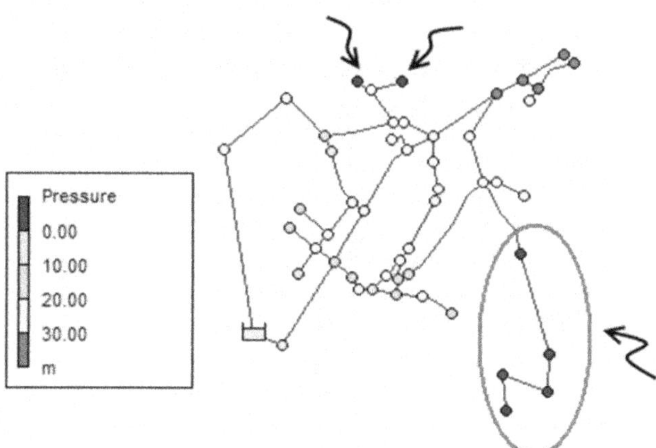

EPANET works using the artifice that negative pressures exist. In reality there are no negative pressures but simply pressure is not enough to get water past a certain point. However when modeling you may have situations, as in the image below, where junction B receives water, even though junction A is dry. In reality if water cannot reach A, it certainly cannot continue to B.

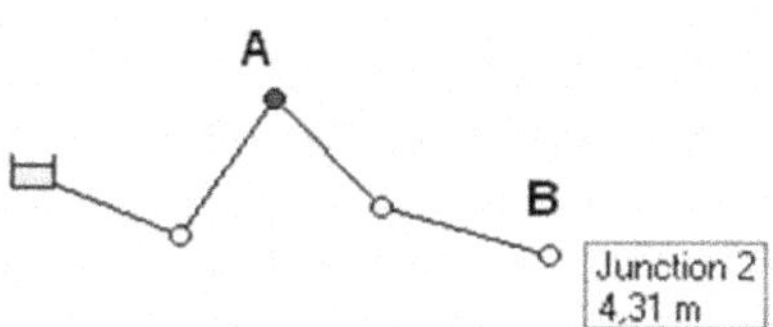

If you get negative pressures and no matter what diameter you enter they stay there, you are probably using coefficients of friction with the wrong formula. If you put values for Hazen-Williams (around 120) when using Darcy Weisbach (around 0.1), you will need enormous diameters. If it is the other way round the same would happen.

https://youtu.be/tPt1Egyk37Y   (Negative pressures in EPANET)

Tracking down invisible errors

We have firstly covered simple errors. The ones we look at now are more dangerous and may go unnoticed leading to a design that in reality is incorrect.

Beginner's luck

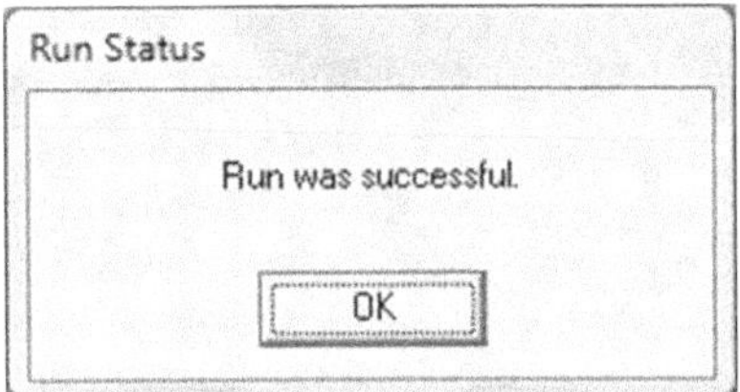

You are tired from working in the field to collect the data, your eyes are stinging from being in front of the screen, you have been getting stressed and feeling pressure to finish, you click the lightning bolt and…. What luck! First time!

This message tastes as sweet as sugar, but will bring you back to down to earth faster than the slap of unrequited love. *Run was successful* can be translated as "the pipes are sufficiently big for the water to arrive at all the junctions with pressure". Sufficiently big goes from the minimum diameter capable of meeting this condition up to infinity. As far as EPANET is concerned, they could have the diameter of the Solar System. After all, it's not going to pay for them!

If you get the message *Run was successful* the next thing you have to do is to track down the pipes that are too big and reduce their size to the minimum ensuring that you keep all points above minimum design pressure.

Modeling without load

Another possibility for having an early *Run was successful* message is the model doesn't have demand. This is the equivalent of putting it in neutral. It has no load and therefore "can cope". The first possibility is that you have forgotten to enter the demand value, that is to say, your junctions have 0 for the Base Demand parameter.

To find out if this is the case, examine the model by clicking *Query*:   

Complete the menu as in the image and click *Submit*. The junctions that meet this condition appear in bold red (grey if you are reading the printed version of this book):

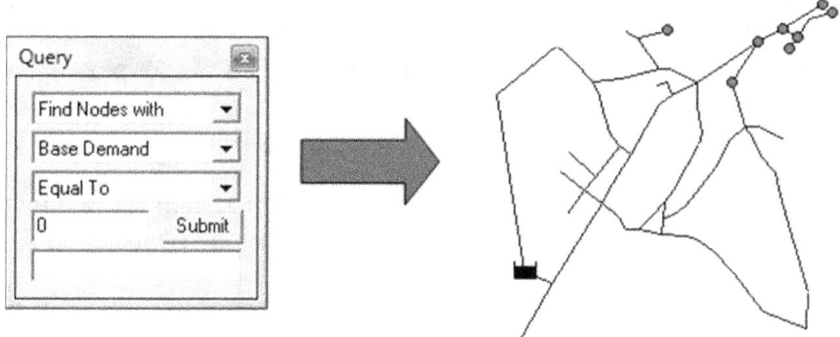

Forgetting to click *Run analysis*

Often there is a phase of small modifications to a correct model or you have a last minute idea. You modify something just as the screen shows the results of the previous calculation and it looks as if it's OK. Always click the lightning bolt to calculate after a batch of modifications. In case you forget to calculate, EPANET will inform you by showing a broken tap at the bottom of the screen. Look closely because it is hardly noticeable!

Others

There are many possibilities, the majority of them due to errors in entering data. Using Query, as in the previous subsection, check:

1. That the points have an elevation by entering "Nodes with Elevation Equal to 0" in the Query box.

2. That the links don't have the default length → "Links with Length Equal to 100" (or the default value that you introduced in *>Project/Defaults/Properties* if you have changed it).

3. That the pipes have a coefficient of friction → "Links with Roughness Above 0".

Visualizing the results of a calculation

The first time you click on calculate, the screen won't change to show any results. You will have to choose what you want to see first. All the viewing options are in the Map tab of the Browser. In the first of the dropdown menus choose the parameter that EPANET should represent in the nodes (or junctions), in the second menu for the links (or pipes), and in the third, the hour of the day.

The buttons **I◄ ◄ ■ ►** allow you to fast forward, rewind and stop the film resulting in superimposing the state at different hours. This will only be visible if you are in extended-period.

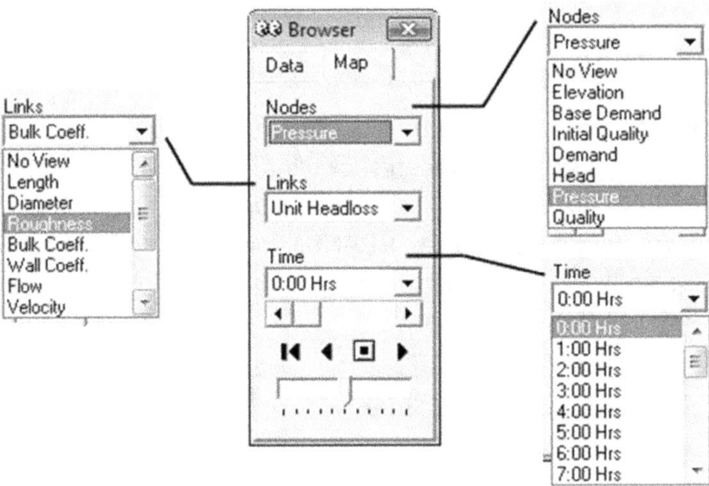

Select one of the criteria we have previously seen. As an example, for unit headloss you would select *Unit Headloss* in the *Links* dropdown menu. You would edit the legend as previously seen in chapter 3, so that it shows the design limits. From here, it's a question of thinking about it and playing with the model to keep all the data inside the design limits.

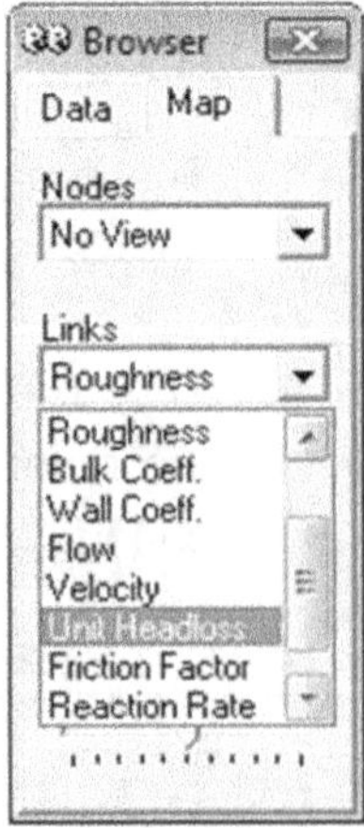

Don't forget to click the lightning bolt if you want to see the result of your modifications. If you change things without clicking it, EPANET cannot take them into account.

Although EPANET can show you the results in many different ways, the color scale is probably the most useful among them. For other forms, like tables or graphs, see the EPANET User's Manual, it explains them very clearly. A very common beginner's mistake is to try to read the pressure in the pipes. Remember that some parameters are visualized in the pipes (i.e. velocity) and others in the junctions (i.e. pressure).

Some recipes for fulfilling the calculation criteria

With practice you will gain experience as to the best way of taking the design to complete the basic calculation criteria. Every network is unique and there isn't one recipe to optimize it. I know very well that a statement like this disheartens beginners who want something concrete to follow. With that in mind and to help speed up the first attempts at the learning process, these recommendations will provide you with the highest probability of success. I trust that you have understood that these recommendations have limitations and are not infallible:

A. Some points exist with pressure less than 1 bar:

1.   Increase the diameter of the pipes that feed them.

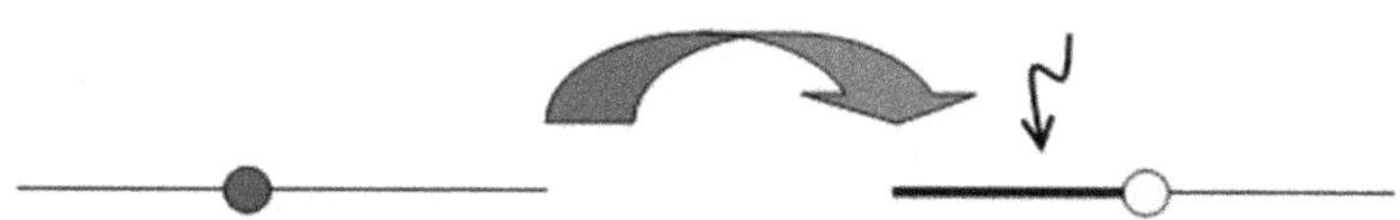

2.   Add new pipes from other points.

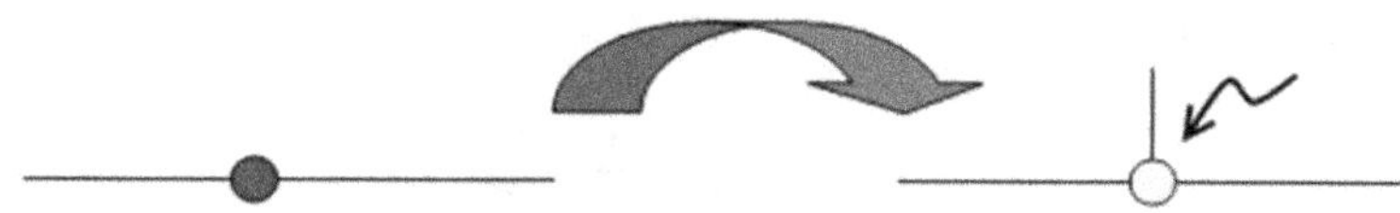

3.   Increase the supply of a big consumer close-by.

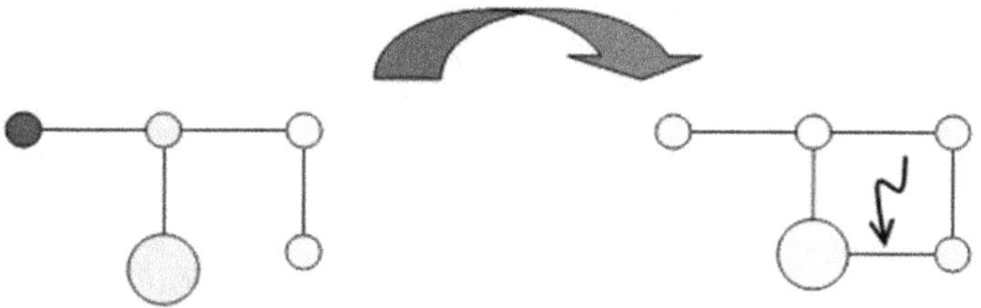

B. A group of points exist with pressure lower than 1 bar:

1. In the proximity of a tank, reservoir by gravity, or pumping station, increase the diameter of the pipes below them. Elevating the tanks should be the last consideration, because of the pumping costs it entails.

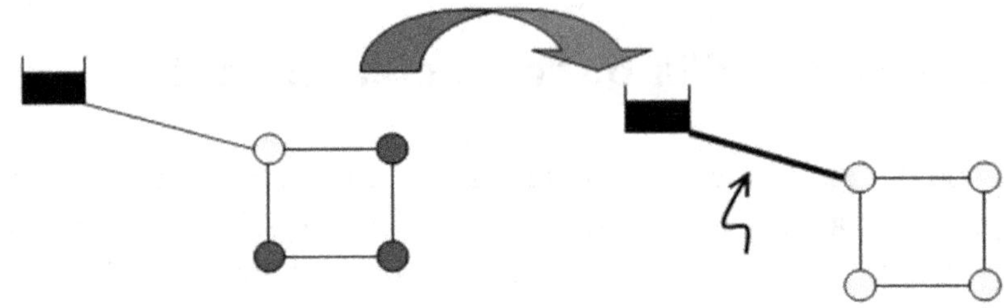

2. In points that are far from a tank, elevated reservoir or pumping station, try to increase the diameter of the pipe that feeds this zone (case A1) or try to put an elevated tail tank.

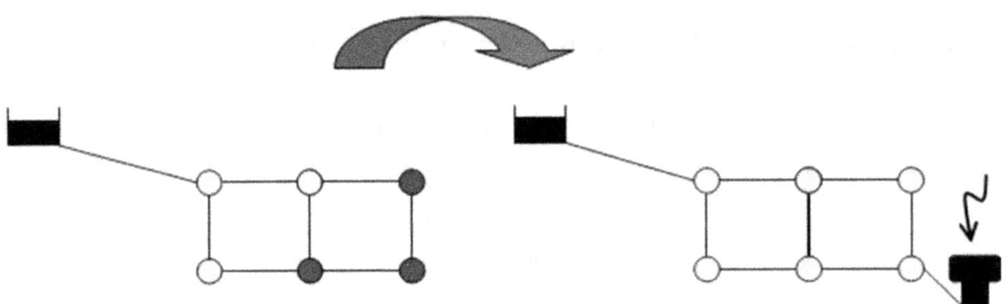

3. If there is a big difference in altitude between two parts of a network, limit the descending paths of the water, reduce diameters or eliminate pipes downstream (alternative to Pressure Sustaining Valve). Consider establishing different pressure zones.

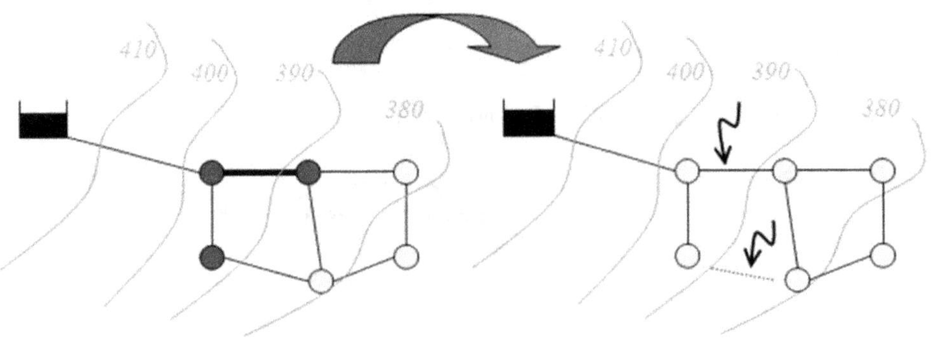

C. Others:

1. If there are velocities above 2 m/s, the pipe is probably too small → Increase the diameter.

2. If there are velocities lower than 0.5 m/s at peak hour, the pipe is probably too big → Reduce the diameter.

3. If the friction is more than 10 m/km, the pipe is too small → Increase the diameter.

4. If the friction is less than 2.5 m/km, the pipeline is perhaps too large → Decrease the diameter.

Pressure is prioritized over the rest of the criteria given. It is what ultimately determines the flow received by the user. Once the pressure is inside the correct margins you can play with the rest of the criteria to get more efficient designs.

Break pressure tanks (BPT)

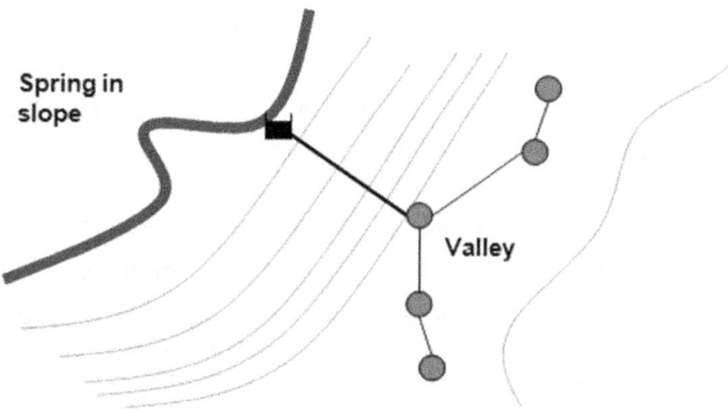

Sometimes, the water source is greatly elevated in respect to the points that it is going to serve therefore the pressure will be excessive. To avoid this, de-pressurize the pipe halfway through with a BPT. However, since you are going to de-pressure the network, the first pipe from the source to the tank can be made very small in diameter and save money on the installation. This implies making it exceed the usual values for velocity and hydraulic gradient. You may consider simply installing all the pipes smaller to avoid excessive pressure however water only loses energy while moving so in hours of low consumption the pressure will still be excessive. This could be the location of the BPT.

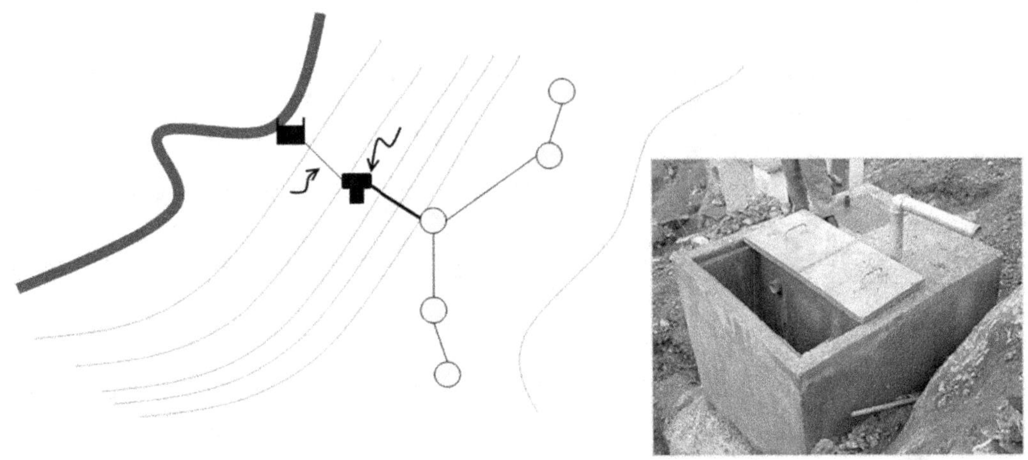

Pressure Zones

Sometimes it is not easy to supply an entire area and have all its users in the correct pressure range. This is a frequent case, for example, in a town on the side of a valley. If the highest users are more than 20 m above the lowest, it is already impossible to supply them all with the 10-30 m range. If the difference is 45 m, for example, when the highest are 10 m high the lowest are 55 m. You have a problem similar to sleeping with a blanket that is too short. Either your feet get cold or your head gets cold!

In addition to the pressure, pressure zones have advantages when it comes to saving on pumping costs as they avoid pumping all the water to the highest zone and save water because from a certain pressure onwards, later increases mean a greater flow from taps and leaks without an appreciable improvement in the level of service. I recommend that you watch this video to understand it in a very visual way:

https://youtu.be/OcKzVfG-pKM (Managing pressure)

Viewing the model in extended mode

Once you have stabilized the model in steady-state, move on to analyzing in extended-period. If you have configured EPANET as suggested in the A way of working section, you simply go to the *Data* tab of the *Browser* and click *Forward*: ▶

Watch how the colors of the objects change with time. If they don't change, it's because either there is no modulation curve defined for consumption or that the ranges of the legend are too big to see the changes.

Confirm that the design parameters are correct, bearing in mind that the most interesting things to observe are:

1.   The periods of least consumption.
2.   The periods of greatest consumption.
3.   The evolution of a parameter with time.

Fire Protection

This refers to the water that is available in case of fire. This is to make sure that at all times there is:

a)   A **fire reserve**, a volume stored exclusively for the purpose. The regulations vary from country to country, but normally is the equivalent of fire flow for two hours.

b)   A **fire flow** that is determined according to the type and number of population.

In practice, the requirements are so large, greater than the peak demand of the population, that the fire flow determines the size of the networks. In development I have seen two approaches, either to completely ignore the protection against fires or to apply the western regulations blindly.

To ignore the necessity of protection against fires is an irresponsible thing to do that deserves no more comment. Applying the western standards or that of the country frequently is disproportionate (in chapter 1 we made the case for a fire flow of 32 liters per second in a community that only has buckets available to fight a fire).

I think that there is a middle point. Determining it is not simple. A good idea in any case is to talk with the local fire department and see what their ideas are, understanding that the firemen are going to participate in the extinguishing of a fire (they don't need to be in white helmets or uniform).

Many times, more than a flow, the important thing is that there is a quantity of water stored to fight a small fire so that it doesn't become catastrophic because the tanks were bone dry.

A fire reserve is created in a normal tank by having the distribution main outlet at a higher point.

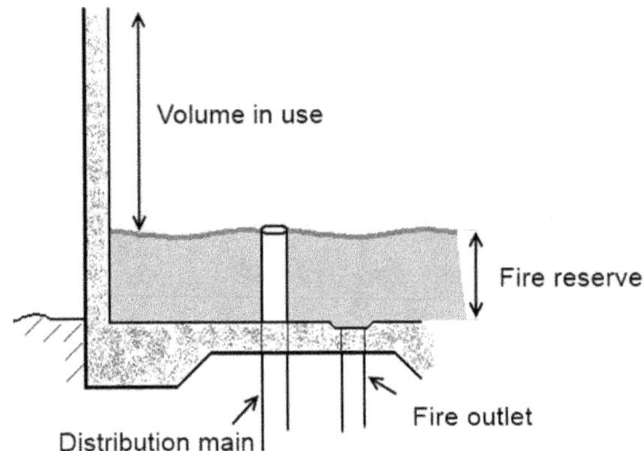

CHAPTER 8

Economic aspects

Introduction to economic evaluation

Economic evaluation tries to answer questions like:

- Which of the alternatives is the cheapest to build?
- Which will be the cheapest to run daily?
- Can the population assume the cost of this operation?
- How will they meet payments to avoid the installations becoming obsolete?

Traditionally, the fundamental objective is:

> **To determine which of the possible alternatives achieves the desired result with the least cost in resources.**

Allow me to add one, which I consider more important in development:

> **Check that the running cost of the alternative is inside the users' willingness to pay** and is therefore **sustainable** once the donors have left.

In practice, this is achieved by comparing the costs of the different alternatives. Each alternative has a capital cost (i.e. buying a car) and another for running costs (insurance, petrol, service ...) In the following section we will see how to calculate them and their limitations however, we must firstly consider a serious issue.

The importance of ethics

In my modest experience, projects seldom carry out an economic analysis. When considering if something works, that is the sole consideration. The button is pressed on the generator and it runs or the pump pumps water. If the generator is running outside its range and is inefficient then that issue is left for the users to resolve.

The economic analysis of the users is implacable. For them a structure that is wasteful doesn't work in the sense that it doesn't improve the living standards. As a consequence, it will be abandoned, many times to the despair of the organization that promoted it who often has claims like that the users "don't maintain anything" or "it is a very difficult context". If the maintenance of the infrastructure is a challenge in itself, the reality is that the absence of economic analysis doesn't help.

In the fairly frequent case of asking for participation from the community it becomes a tricky issue. Here there is the potential to have a frankly negative impact through decapitalization. The users placed their trust in the technical capacity of the assistance, and their contribution will be linked to results. If like the San fish factory in Mali, each kilo of fish ends up costing 4,000 dollars[13], despite the work, effort and commitment of many people the conditions do worsen.

Common nonsense

Apart from the complete absence of economic evaluation, the three most common blunders with respect to economic issues are:

1. **"Granny Economics"**. This refers to the obsession to save by proposing activities that in reality are a waste of effort and money, compromising the overall results and disheartening and disheartening all involved. With this approach the most important thing is to go scrape money from wherever you can to make the costs small.

2. **"Economic Despotism"**. This refers to the ideal that everything is determined economically but to avoid controversy, let's say our capacity to measure it is severely limited. For example, what is the value of educating someone? Does a dollar really have the same value for a middle-class westerner as to a person below the poverty line? Yet both dollars buy the same amount of potatoes. What is the flaw? To be fanatically guided by aspects of pure efficiency is inconsiderate in the majority of cases. The value to the people is obviously what really matters.

3. **Vague calculations of hazy ideas.** This refers to calculating and budgeting without clear definition. To calculate the costs you need a clear idea in your head of what components are required, so many km of pipe, so many valves of a certain type, so many truckloads of sand... To calculate the total cost of a water network without precise detail is like calculating how much *a machine* costs... a machine of what kind?? The design precedes the budget and never the other way round, as occurs

---

[13] *Handcock G. 1989. "Lords of Poverty"*

so often in development. Approaches such as "we have 200,000 dollars I'd reckon that is enough for a network", betray and spoil the results.

Determining the willingness to pay

We have seen that the fundamental objective in economic evaluation is to check that running costs of the alternatives are within the users' willingness to pay and therefore is sustainable once the donors have left.

That means the decision as to whether the capital and running costs are acceptable doesn't fall on the person designing but on the users. It's therefore vitally important that the decision maker can communicate properly with the future users.

What the users are willing to pay may be tricky to determine since they may not have favorable experiences from the past, or never had a proper system before. As a guide, the cost should not exceed 3-5% of the family income. Sometimes information is as easy to get as establishing the cost of the traditional systems:

The techniques to work out how much a person would pay fall outside the scope of this book, but a good place to start is *Willingness-to-pay Surveys* published by WEDC in English and downloadable for free on their site. Next we see how to assess the costs the users will bear.

Determining the capital costs' bill

This is the most obvious cost for the equipment and installation and because it is paid all at once is the largest component of the project's budget. It is also the one which commands the most attention.

Repayment and devaluation of money over time

Although the capital cost is paid all at once, it can be distributed over the whole lifetime of the installation. One of the first decisions to be taken is to determine the **repayment period**, in other words, how long we expect to use what we are buying and constructing. This is not an easy decision. How durable is a water network? How durable is a house? To complicate things, the durability of a house, for example, depends on the maintenance costs that we are willing to assume, costs that don't form part of this bill.

Don't get overwhelmed by this. It's important to remember that economic evaluation is an approximate method and that there are always some reasonable criteria to follow. For example, in the case of a water network, this repayment period should be at least equal to the design life. If it is designed for 30 years, the repayment period should be 30 years and to obtain the corresponding annual cost divide the actual capital cost by the number of years.

However there are other phenomena to take into account, mainly, that **the value of money diminishes with time.** In 1950 a bus ticket cost two cents now it costs $3. Why? Because to compensate the loss in value of that initial two cents, the price of the ticket has had to be increased each year. The initial two cents eventually arrives at $3 after 50 years. These costs are usually compared and taken into account at the start of the project.

Obtaining the annual cost of a capital investment

Have a quick look at the theory which can be expanded in any economic analysis book; the way to proceed is as follows:

1. Work out the **interest, i**, that a bank would give you to deposit a similar amount and convert it into a decimal. For example, 3% → i = 0.03.

2. Predict what the **inflation, s,** will be for the period considered. You can look at a few years in the World Bank data and make a guess, because it's not possible to know how it is going to develop in the future. This will be your parameter **s**, also as a decimal.

3. Calculate the **real interest rate, r**. This rate takes into account interest and inflation. If inflation is more than the interest a bank would give you, the money is worth more in the present than it would be in the future. If they are equal, it will maintain the same value and if the bank interest is more than inflation, the value of the money will increase. It is calculated through this equation:

$$r = \frac{1+i}{1+s} - 1$$

4. Calculate the **repayment factor a_t** for t years:

$$a_t = \frac{(1+r)^t * r}{(1+r)^t - 1}$$

5. The annual cost of the capital investment is the invested amount, M, multiplied by the repayment factor:

$$F = M * a_t$$

Example

A network for water supply is budgeted at 100,000 € and designed for 30 years is evaluated in Uzastan where the banks give 5% interest and the average inflation for the last few years has been 4.5%.

The interest is i = 0.05 and the inflation is 4.5%, so s = 0.045.

The real interest rate is:

$$r = \frac{1+i}{1+s} - 1 = \frac{1+0.05}{1+0.045} - 1 = 0.00478$$

The repayment factor is:

$$a_t = \frac{(1+r)^t * r}{(1+r)^t - 1} == \frac{(1+0.00478)^{30} * 0.00478}{(1+0.00478)^{30} - 1} = 0.03586$$

The annual cost is F = 100,000€ * 0.03586 years^{-1} = 3586.24 €/years, approximately 3586 €/year.

Notice that it's different to 100,000€ / 30 years =3333.33 €/year. That is due to the corrected value for the capital cost, called **present value**, is F * 30 years = 107,587€ and not the 100,000€ of the capital cost.

Determining the running costs' bill

In water systems that aren't gravity-fed the main cost is energy or fuel for the pump, followed probably by the treatment of the water. Conceptually this cost is much simpler to produce; it is an inventory of all the expenses that the network will incur in one year of working. Saying this, there are some costs that are very evasive, for example, breakdown repair. In development, rarely the economic decisions are going to be so accurate. The servicing expenses are comparatively small in correctly designed systems and usually there are other criteria that prevail in narrow margins.

Pumping expenses

In many instances it is not worth involving EPANET in the calculation of pumping expenses. They are easier to calculate by hand with this equation:

$$E(kWh) = \frac{mgh}{3.6 * 10^6 \; \eta}$$

Where:
  m, is the mass in kg
  h, is the total pumping head
  g is 9.8 m/s²
  η, is wire to water efficiency, 0.6 is a common value.

Look at the following example.

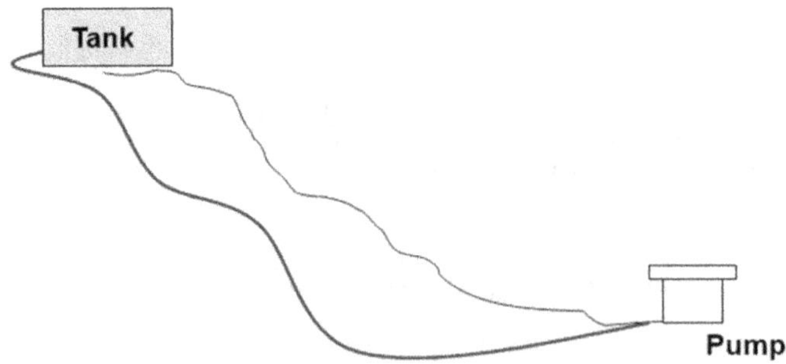

A pumping station fills a tank that supplies the city. The population supplied is 1,000 people and it has been established that each resident will receive 50 liters a day. The price of a kWh is 0.40€ and doesn't vary during the day. The total pumping head is 70 m.

The annual consumption is: 365 days * 1000 per * 50 l/per*day * 1 m³/1000 liters = 18250 m³.

$$E(kWh) = \frac{mgh}{3.6 * 10^6 \; \eta} = \frac{18250000 kg * 9.8 m/s^2 * 70m}{3.6 * 10^6 \; * 0.6} = 5796.06 kWh$$

And the annual cost will be: 5796.06 kWh/yr × 0.4 €/kWh = 2318.43 €/yr

Often the electrical energy is generated on site with a generator. In this case, the cost of a kWh will be the price of diesel. The consumption of an average generator is 0.35 liters of diesel per kWh produced.

In the previous example, if the price of diesel is 1 €/ liter:

5796.06 kWh/yr × 0.35 l/kWh × 1 €/l = 2028.62 €/yr

Knowing the cost per hour of the pump you can proceed in the same way.

When to use EPANET?

In all situations when the calculations become difficult but above all:

- Pumping directly to the network. The pressure that the pump encounters changes according to the hourly consumption and therefore the flow and price /m³ of water.

- Many pumps, even if their workload is regular.

- Automatic start/stop pumps or variable velocity pumps, both uncommon in development.

Bill comparison and contextualizing

You have devised three alternative interventions A, B and C. For each of these you have established the annual capital cost and the running costs. The only thing that you have to do now is to compare and contextualize them. The comparison is simple; the smallest of them is A.

Type of cost	Alternative A	Alternative B	Alternative C
Capital	8000	10000	7000
Running	3000	6000	4500
TOTAL	11000	16000	11500

Contextualizing through ponderation

That alternative A is the cheapest doesn't mean that it is the most desirable. Other types of criteria exist; for example, in an emergency situation the speed of construction will be put above the construction cost. To take into account all the decision criteria called "contextualizing a cost", a ponderation must be made.

Ponderations are done because people have serious problems to manage more than 4 or 5 variables in the head. Imagine for a moment that someone asks you what you want to eat while you try to figure out the number of days between the 8th of January and 23rd February… As a consequence of that, we have a strong inclination to decide using only one parameter and ignoring the rest: "Huuummmm…, the same as you", I bet you would respond.

To maintain objectivity and to document the decision making process, each variable is analyzed independently. The process is very simple. For each criterion (speed, economy, gender approach…) a weighting according to its importance is given. In the same way, for each alternative (wells, water truck, gravity system…) give another weighting with respect to each criteria. The alternative should score highest points when it contributes something to the users by its abundance/presence or by its absence/shortage, in other words, it scores more when it has a better impact.

Suppose that the pumping station of a small riverside city has broken down due to a flood and I am trying to decide between repairing the station or hiring water trucks from a neighboring town (I don't have the funds to do both). In this case speed is fundamental so I weight it 9 out of 10. I have sufficient funds for both alternatives so that the economics of the solution is not so important. I weight it 4 out of 10. The implementation speed of water trucking is very much needed, and scores 10. The cost is much higher, I give it a 0. Rebuilding the station is a very lengthy process because there are parts to be ordered from abroad, it scores 1. However, it's very economical as it is only the electrical parts that have broken. It scores 8.

(A) Water truck: **(B) Repair:**

Speed   9 * 9 = 81   Speed   1 * 9 = 9
Cost   0 * 4 = 0   Cost   8 * 4 = 32
            TOTAL   81   **TOTAL   41**

The score for the water trucks is higher and it is decided to implement this alternative.

Disappointed? Wasn't the decision obvious anyway? Sure, but this was a very simple example. This is a real one from Tanzania:

| | Pump (A) | | Gravity (B) | | |
Criteria	Score	Weighted score	Score	Weighted score	Weight
Quantity of water produced	2	20	9	90	10
Water quality	5	40	7	56	8
Low risk of water trucking	1	8	9	72	8
Damage by returning refugees	9	63	2	14	7
Low risk of disturbance	1	7	10	70	7
Storage	1	7	9	63	7
Diversity of sources	2	12	8	48	6
Gender	3	18	8	48	6
Increase in population served	0	0	8	40	5
Donation of funds	4	20	7	35	5
Economic analysis	7	28	4	16	4
Social benefits	2	6	5	15	3
Implementation time	8	24	4	12	3
Organizational demand	2	4	4	8	2
Complexity	8	16	5	10	2
Technical risk	7	14	8	16	2
TOTALS	**Pump**	**295**	**Gravity**	**613**	

Limitations and sources of uncertainty

Throughout the explanation you will have picked up on some limitations with the way of working. These are some:

1. We don't know what inflation will be in the future.

2. The lifetime of equipment varies a lot from one unit to another. Although the average life of a car could be ten years, some cars with the same specification will last longer and others less.

3. The average life is not clear, like the life of a person. It is difficult to determine when the service period of a water network will end. To a great extent it is a decision that depends on economic criteria. It will be abandoned when the maintenance costs become excessive, but that depends on the available alternatives and the costs in the long term.

4. There are expenses that are very difficult to determine. For example, servicing.

5. The stability of prices is unpredictable. Diesel can triple, skilled labor becomes more expensive with development, or pipes can become cheaper with the price of oil…

…

This is about arriving at a sufficiently good approximation, despite the uncertainties, to be able to base a decision on it. After all, no decision is taken completely informed; there is always a good number of unknowns. If you keep this idea of "sufficient approximation" in mind, you will see that limitations are not that restrictive.

Using EPANET to budget

What we want is for EPANET to break down how many meters of each type of pipe rather than having to work it out one by one by counting. Unfortunately EPANET's export options are a little limited.

1.   Click the icon *Table* , so that this dialog box appears and select *Network Links*. In the *Columns tab*, tick the boxes besides diameter and length.

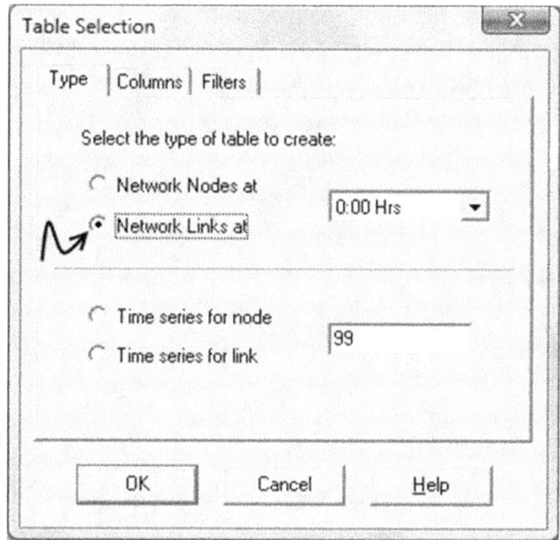

2. Click accept to generate a table similar to this:

Link ID	Flow LPS	Velocity m/s	Unit Headloss m/km	Friction Factor	Reac m
Pipe 497.1	0.69	0.04	0.02	0.044	
Pipe 497.5	-1.82	0.10	0.10	0.029	
Pipe 497.7	-2.31	0.13	0.16	0.028	
Pipe 497.8	-2.49	0.14	0.19	0.028	
Pipe 497.9	-2.67	0.15	0.21	0.027	
Pipe 497.10	2.96	0.16	0.24	0.027	

3. Select the *Diameter* and *Length* columns clicking above the first box and dragging sideways and downwards until it includes all the pipes.

Link ID	Length m	Diameter mm
Pipe 497.1	8.86	200
Pipe 497.5	67.	200
Pipe 497.7	151.4	200
Pipe 497.8	182.4	200
Pipe 497.9	60.9	200
Pipe 497.9	81.67	200

4. To copy you have to go to the menu *>Edit/Copy To* (Control C and similar don't work). It displays this box and you should click accept.

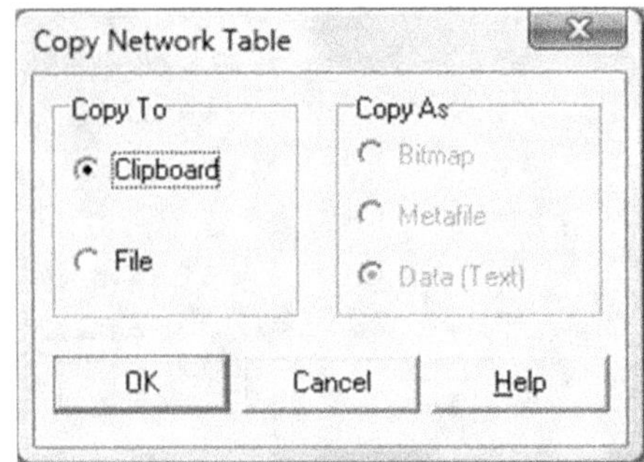

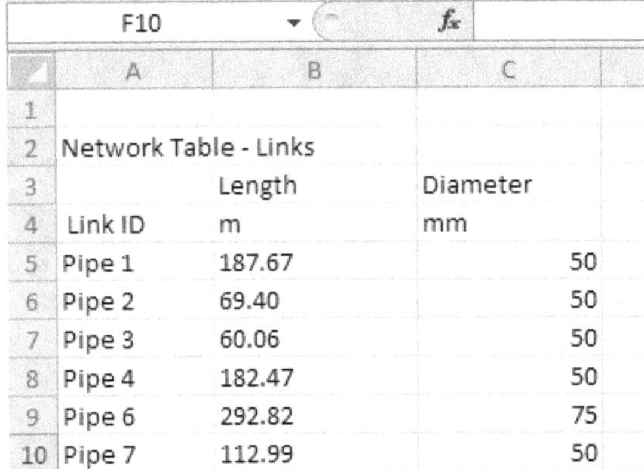

5. Copy your data into a spreadsheet:

6. In Excel, follow the path *>Data/Sort*. In the menu choose Diameter. This will put the pipe with the same diameter together so that you can calculate the number of meters for each diameter.

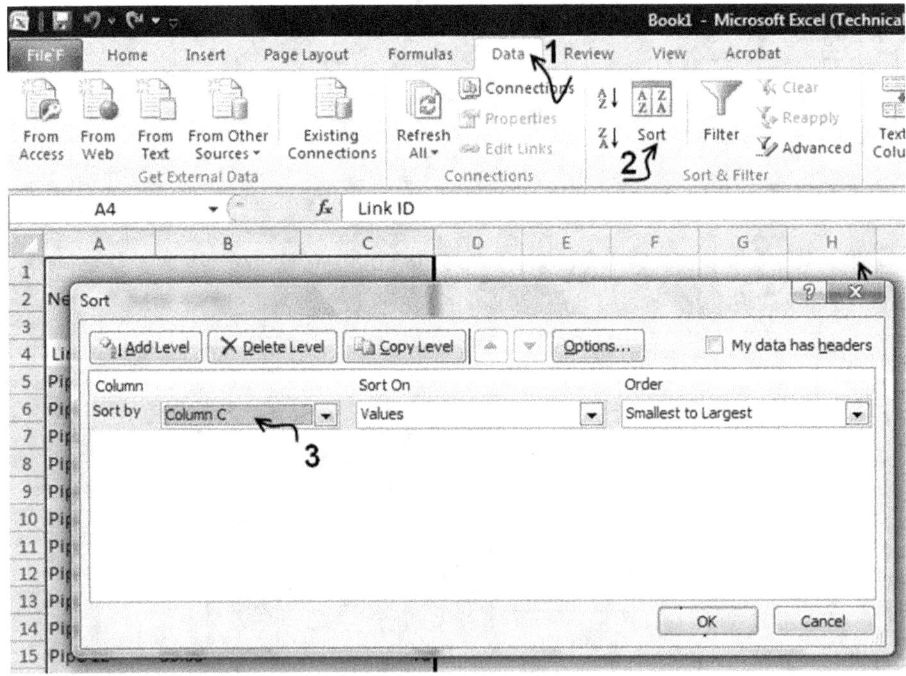

7. Calculate the totals for each diameter and assign a price per meter of line that includes excavation, backfilling, installation, etc.

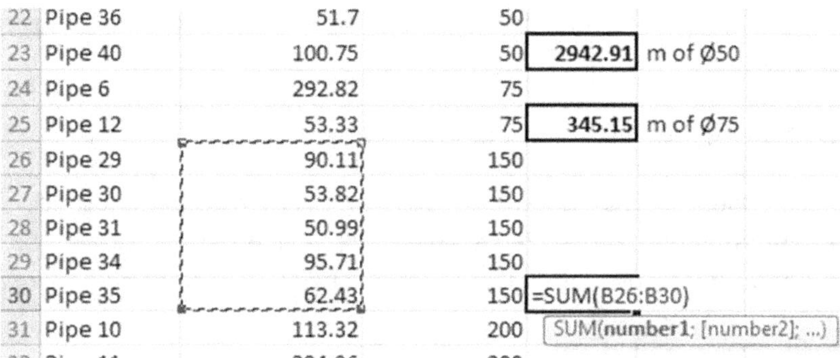

22	Pipe 36	51.7	50		
23	Pipe 40	100.75	50	2942.91	m of Ø50
24	Pipe 6	292.82	75		
25	Pipe 12	53.33	75	345.15	m of Ø75
26	Pipe 29	90.11	150		
27	Pipe 30	53.82	150		
28	Pipe 31	50.99	150		
29	Pipe 34	95.71	150		
30	Pipe 35	62.43	150	=SUM(B26:B30)	
31	Pipe 10	113.32	200	SUM(**number1**; [number2]; ...)	

From here it's in your own hands as it would leave the purpose of this manual to continue with the use of a spreadsheet or procedures to make a budget.

Using EPANET to determine energy consumption

We have seen that in the majority of cases it is easier and less prone to error to calculate by hand. However if you are in a situation where it will be laborious and complicated to do it by hand, you can get an energy report that looks like this:

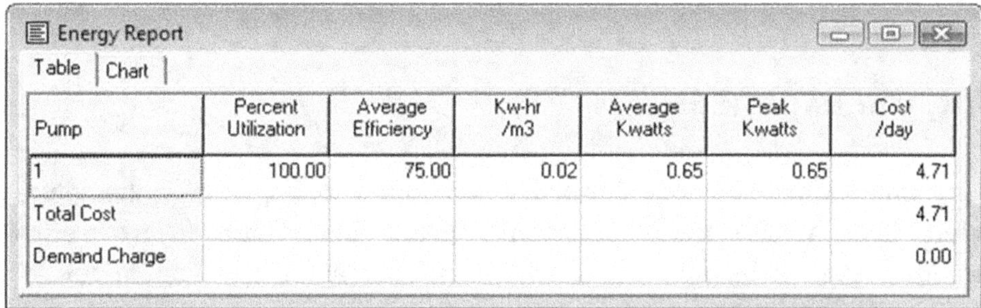

Pump	Percent Utilization	Average Efficiency	Kw-hr /m3	Average Kwatts	Peak Kwatts	Cost /day
1	100.00	75.00	0.02	0.65	0.65	4.71
Total Cost						4.71
Demand Charge						0.00

To get it, follow these steps:

1. Determine the energy price and if it has variations. Many countries have an energy surplus at night making electricity cheaper.

2. In the *Data* tab of the *Browser*, select *Options* at the bottom of the dropdown menu and *Energy* afterwards, to display this dialog box:

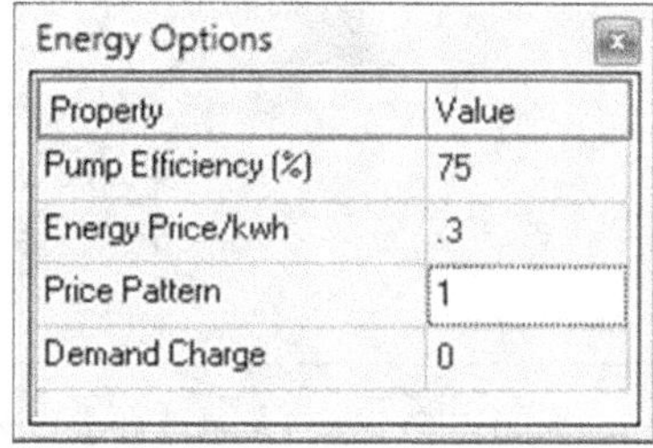

3. Enter the price of a kWh. If the energy will be provided by a generator, assume the 30% price of a liter of diesel as the cost of a kWh. If there are hourly differences in price create a modulation curve as we have seen in sections of chapter 4. The *Demand Charge* is used if there is a charge for the maximum power used, which is rarely the case.

4. To get the energy report, follow the path *>Report/Energy*.

If you want to split it finer, you can enter a modulation curve for the output. You will get this curve from the pump manufacturer; it is expressed in % and can be labeled as η. It is the overall efficiency that is also called the wire-water efficiency. It is created in the *Curve Editor* as explained in chapter 4 for the volume curves. As a reminder:

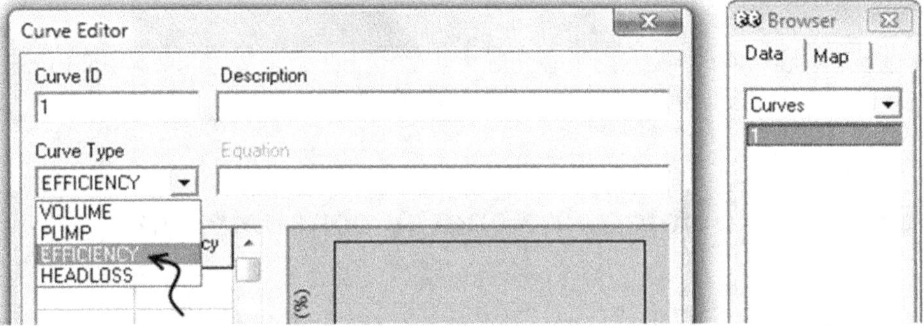

To enter it click on the pump in question and go to the property *Curve type* and select *Efficiency*.

Ranking the expenses

How much money is spent where in building a water network depends on the nature of the infrastructure. Some authors, like Stephenson[14], make a generic breakdown to distribute the costs like this: 55% for buying the pipes, 25% excavation and installation. In development, excluding the organizational costs of the NGO and of access to water (boreholes, etc.), the following figures are closer to my experience:

1°.	Pipes and accessories	36%
2°.	Excavations	31%
3°.	Layer of sand	16%
4°.	Valve boxes	11%
5°.	Pipe installation	5%

---

[14] *Stephenson (1981) "Pipeline Design for Water Engineers".*

Here we can draw some interesting conclusions. Points 2, 3, 4 and 5 can be considered relatively independent to the pipe diameter. Thus, two thirds of the capital cost is independent of the pipe diameter.

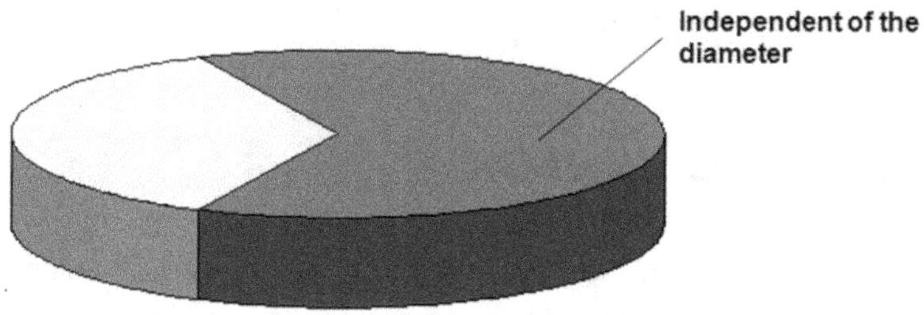

The conclusion is so important it deserves its own section.

"Dry diametritis"

Clearly this is not an engineering term but a tactic to help you remember this illness that many networks suffer from in development. The outcome generally is because of a granny economics approach or of a restriction *it has to cost less than x.*

Basically you try to save on the pipes by limiting the diameters to the very lowest. The resulting networks barely tolerate design errors or changes of use, are not easily enlarged, let down the users at the time of day when they most need water and are costly to operate. It is not surprising that these networks are frequently dry. If this has been accompanied by savings in the sad bed for the pipe and the pipes laid in shallow excavations resulting in frequent bursting the result doesn't leave much cause for celebrations.

To duplicate a line because it isn't capable of transporting sufficient flow is very expensive. Let's look at an example comparing 1,000 meters of a pipe of 200 mm, with two of 160 and 125 mm that transport the same amount of water. Note that 160 to 200 m is only one step in diameter:

	200 mm	**160 + 125 mm**
Pipe price	21,000	13,500 + 8,100
Installation etc. (64%)	37,300	37,300 + 37,300
TOTAL	**58,300 €**	**96,200 €**

Therefore, **be generous with the diameter of pipes when they are plastic** and you will enable the population served to be able to:

1. Postpone enlargements.
2. Facilitate enlargements.
3. Radically lower the pumping costs.

Be generous… but how much and how?

Being generous with the diameters has two main side effects, a worsening of quality by increasing the residence time in the network and an increase in costs.

The best candidates to promote are:
   (1) Downstream of tanks or pumping lines.
   (2) Pipes that form part of a grid.
   (3) Important pipes that could form a grid in the future.
   (4) Pipes that go to zones of possible development.

In the image on the right, possible candidates are in bold.

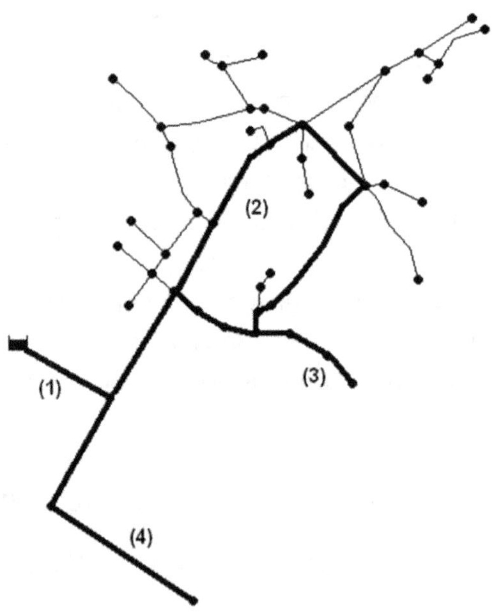

The next question is how much should diameters be increased. There is a diameter below which there is a spectacular decrease of the water transport capacity. Take two diameters larger than the failed size, that is to say, one size larger than required.

The illustration below represents the development of pressure of a pipe through time, assuming initially that it is 4" and increasing to 8". Each line on the graph represents a pipe diameter. The pressure interval has been exaggerated to make it easier to see:

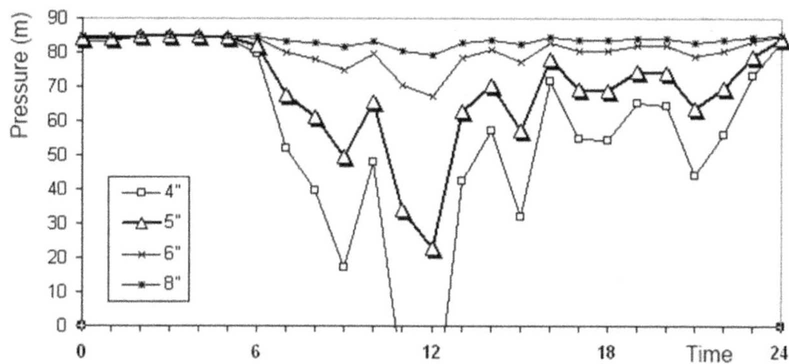

The first pipe that maintains the pressure is in bold, and corresponds to 5" (triangles). According to the recommended approach, the pipe to install would be the next available commercial diameter, 6". Look at hour 12 in the day and see that bigger diameters are very similar. In the majority of cases the practice of passing a diameter allows enlargements of the network without the need to install new pipes.

Cost vs. Diameter

Pipes and transport capacity

It is often assumed that bigger pipes transport the l/s at a smaller price. The logic behind this is that if you bulk buy 300 pens the cost per unit should be cheaper. However, for plastic pipes that cannot be taken for granted. The cost to carrying capacity ratio is almost constant.

The two graphs below illustrate this for PVC (Uralita) and HDPE (Chresky), approximately 1 €/m for each l/s in PVC and 1.1 €/m for the HDPE. The per-liter cost is the thick horizontal line close to the axis:

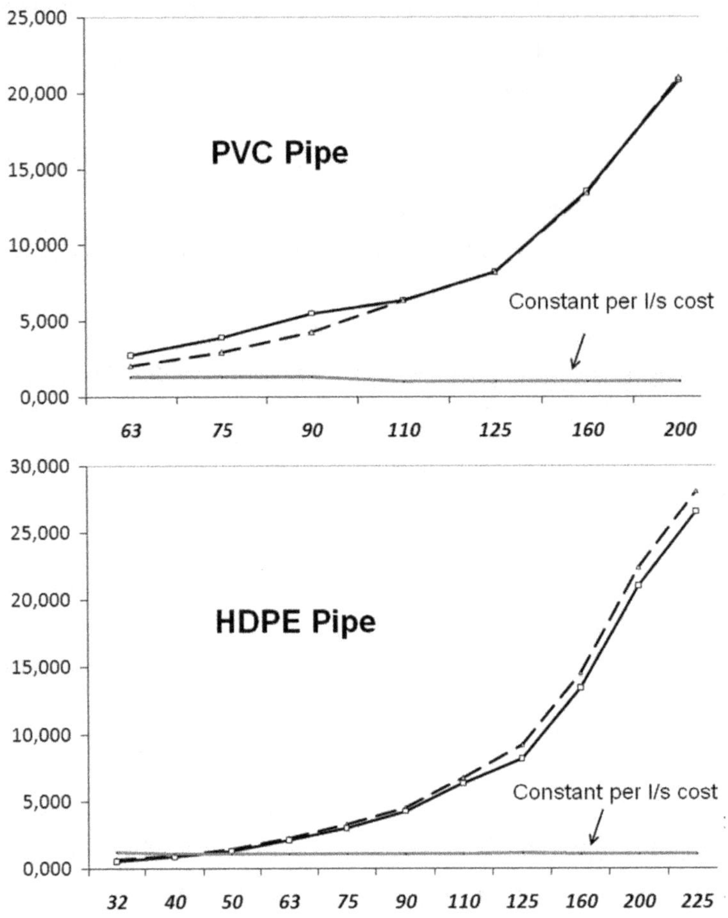

In both cases the transport capacity of the pipe in l/s (solid line) and the cost per meter (dotted line) increase parallel with the increase in diameter.

Accessories

An important problem is that the price of accessories, notably the valves, shoots up disproportionately with diameter. If a gate valve costs 11 dollars for 1", it costs 1,460 for 12". You can see the relationship of price against diameter in the graph that follows.

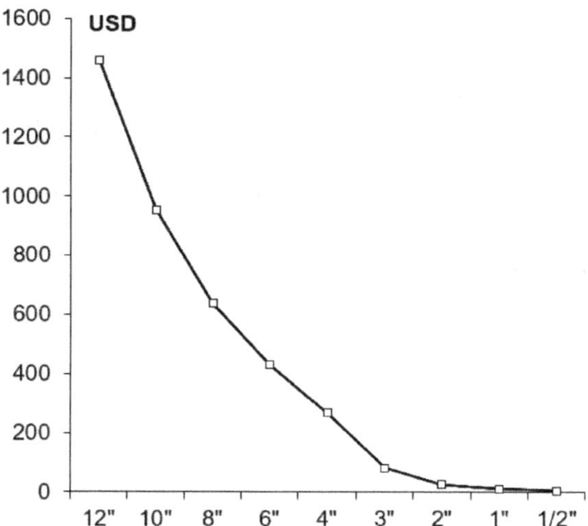

The bigger the unit price of an item, the lower the probability that it will be replaced by the operators or the community. Although it may not feel like a big deal to you, especially compared to the total capital cost of the project, take into account that 700 USD in some instances are many days pay. The ultimate consequence is that, **replacing control accessories for larger diameters can present a big challenge at the community level**.

Often, at first sight the network seems to work as it is common for communities not to notice a failed fitting. It's not as evident as a broken pump which stops supplying water. Nevertheless, if a gate valve breaks in the process of opening or closing, for example, they will remain half closed and the transport capacity of the pipes diminishes significantly.

Most common wasteful designs

As we are coming to the end of this introductory book you probably have a model loaded, you have worked it and it is ready to be implemented. Before you rush to implement it in the heat of the moment, finally check that you are not in one of the following situations:

Strangling of mains

This happens in a network where pipes are too small at the exit from a tank or pump. The result is a drop in overall pressure in all points which can be "resolved" by putting the tanks at a higher elevation or choosing a more powerful pump. This is like accelerating a car while pressing the brake. The fuel expenses shoot up. To solve this, increase the diameter of these pipes and lower the tanks and/or reduce the power of the pumps.

Gigantism

This is a case that often goes unnoticed because the network works fantastically. It comes as a result of laying pipes much bigger than necessary which damages the quality and increases the capital and maintenance costs. The simplest way to detect this is to look at the velocity the pipes have in the period of maximum consumption. If it is less than 0.2 m/s, it is a suspect of gigantism although it's not always the case.

Redundancy

This consists of laying pipes that contribute nothing to the transport capacity in places where geographically it is unnecessary. We have seen the cost of installing a pipe parallel to an existing one that lacks sufficient capacity. To install redundant pipes is very similar.

Remember the data:

	200 mm	**160 + 125 mm**
Pipe price	21000	13500 + 8100
Installations etc. (64%)	37300	37300 + 37300
TOTAL	***58300 €***	***96200 €***

It is evident that to put pipes in practically the same trench is very similar to either of these two diagrams.

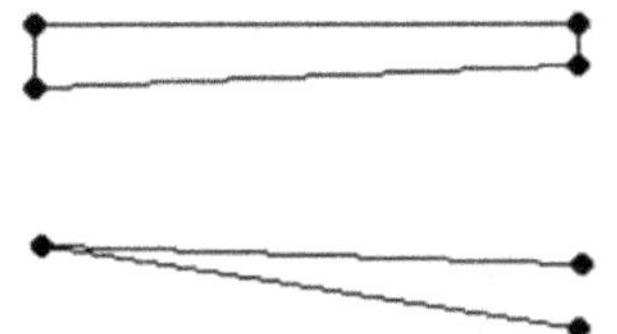

We have a strong tendency to draw networks similar to this because you will see them frequently in books.

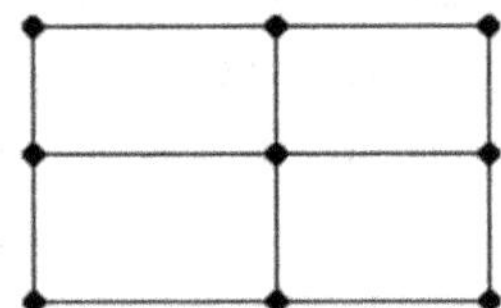

Would this not be equivalent in many cases?

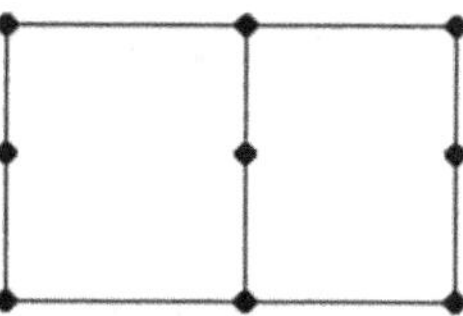

Even this, providing the terminal distribution does not raise a problem, for example with the quality.

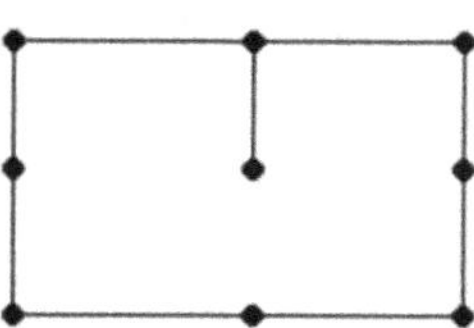

One of the best ways to avoid this is by making use of simplification, similar to skeletonization, but removing the actual pipes and not only in the model.

Bibliography

1. Arnalich, S. (2010). *EPANET and Development. A progressive 44 exercise workbook.* Arnalich, water and habitat.   www.arnalich.com/en/books.html

2. Cabrera E. et al. (2005). *Análisis, Diseño, Operación y Gestión de Redes de Agua con EPANET.* Editorial Instituto Tecnológico del Agua.

3. Expert Committee (1999). *Manual on Water Supply and Treatment.* Government of India.

4. Fuertes, V. S. et al. (2002). *Modelación y Diseño de Redes de Abastecimiento de Agua.* Servicio de Publicación de la Universidad Politécnica de Valencia.

5. Mays L. W. (1999). *Water Distribution Systems Handbook.* McGraw-Hill Press.

6. Santosh Kumar Garg (2003). *Water Supply Engineering.* 14º ed. Khanna Publishers.

7. Rossman, L. (2000). *EPANET 2 Users Manual.* Environmental Protection Agency. Cincinnati, USA.

8. Walski, T. M. et al. (2003). *Advanced water distribution modeling and management.* Haestad Press, USA. Haestad methods.

9. Walski, T. M. wt al. (2004). *Computer Applications in Hydraulic Engineering.* Haestad Press, USA. Haestad methods.

www.ingramcontent.com/pod-product-compliance
Lightning Source LLC
La Vergne TN
LVHW080431200726
843507LV00004B/787